CONTENTS

IN THE AIR	6 — 7
Afloat in the air: AIRSHIP	8 — 9
Airborne destroyer: FLYING FORTRESS	10 — 11
Giant cargo plane: GALAXY	12 — 13
Transatlantic giant: JUMBO JET	14 — 15
Whirling wings: HELICOPTER	16 — 17
Hover power: HARRIER JUMP JET	18 — 19
Supersonic airliner: CONCORDE	20 — 21
Man into space: APOLLO	22 — 23
Moon landing: LUNAR MODULE	24 — 25
ON THE GROUND	26 — 27
For safety's sake: FAMILY CAR	28 — 29
Super bus: GREYHOUND	30 — 31
Fast racer: FORMULA ONE CAR	32 — 33

LOOKING INSIDE
MACHINES ON THE MOVE

DAVID SHARP

Rand McNally & Company Chicago • New York • San Francisco

Space-age giant: CRAWLER TRANSPORTER	34 — 35
Gathering the grain: COMBINE	36 — 37
The iron horse: CASTLE CLASS STEAM ENGINE	38 — 39
On the rails: DIESEL AND ELECTRIC LOCOMOTIVES	40 — 41
Inter-city: HIGH SPEED TRAIN	42 — 43
Armored to battle: MARK IV AND SHERMAN TANKS	44 — 45
Mechanical digger: POWER SHOVEL	46 — 47
ON THE WATER	48 — 49
Cars across the sea: CAR FERRY	50 — 51
Boxed cargo: CONTAINER SHIP	52 — 53
Last of a line: QUEEN ELIZABETH II	54 — 55
Boat on skis: HYDROFOIL	56 — 57
Land and sea skimmer: HOVERCRAFT	58 — 59
Deep sea worker: PISCES	60 — 61

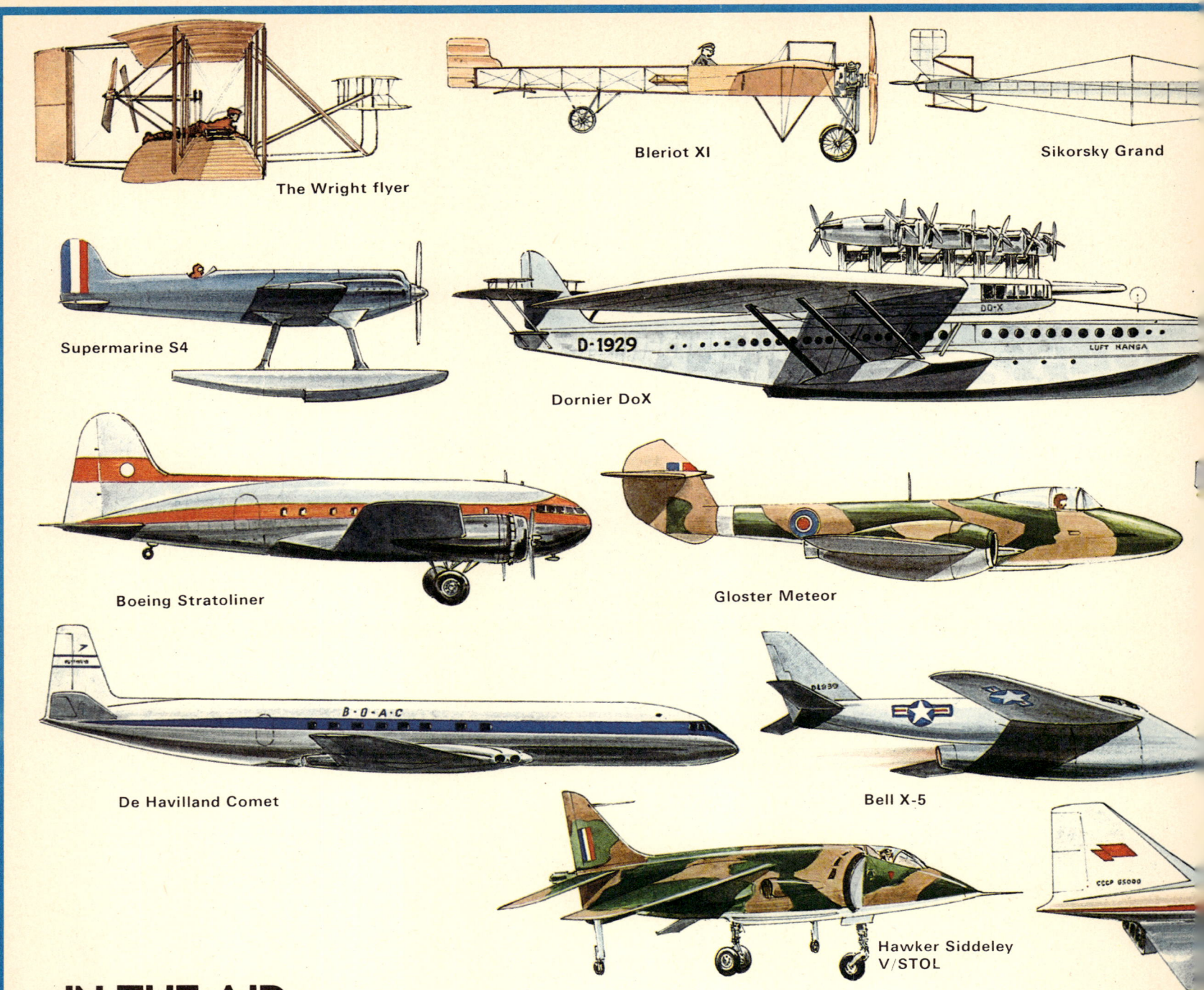

IN THE AIR

After centuries of hopeful theorizing and experiment, Jean-Francois Pilatre de Rozier rose skywards in a tethered Montgolfier hot air balloon in October 1783. He remained airborne for an exhilarating 4 hours 24 seconds. While this feat can hardly be called flight, it was an early step on the great adventure that led to the first powered flight in an aircraft by the Wright brothers in 1903. Just before the end of the year, on the 31 December, Orville Wright flew 120 feet (36.6 meters) in 12 seconds. This epoch-making success was the product of decades of hard effort by pioneers of flight since Sir George Cayley browbeat his petrified coachman into piloting the first manned glider in 1853.

French aviators quickly challenged the American lead in powered flight. Louis Bleriot captured public imagination by flying across the English channel in his Bleriot XI on 25 July 1909. His machine achieved speeds up to 36 mph (58 kmph).

Igor Sikorsky produced what was by far the largest plane in the world in his four-engined 'Grand', which first flew in 1913. The open cockpit was, Sikorsky declared, necessary to the pilot's swift awareness of any change in the aircraft's angle of flight – a clear example of pilots' claims "to fly by the seats of their pants."

During World War I, military interest brought a great improvement in airplane engines. At the start of the war, the Gnome engine was efficient. It weighed 172 lb (78.05 kg) and produced 50 hp. By the end of the war, four of the 375 hp Rolls-Royce Eagle VIII engines mounted in the Handley-Page V/1500 brought Berlin within bombing range of bases in England.

The Vickers Vimy bomber was not completed in time for war service, but in a converted version, Captain J. Alcock and Lieutenant A. Whitten Brown flew from Newfoundland to Great Britain to make the first direct crossing of the Atlantic by air. They crash-landed in foul weather in Ireland on 14 June 1914, having flown at an average speed of 118 mph (189.9 kmph).

Experimental aircraft, such as the Cierva autogiro, led to improved stability and safety. The comfort of passengers was beginning to occupy the thoughts of airline owners. Most of their machines were converted World War I bombers but by 1922 factories were producing aircraft purpose-built for passenger services.

Speed, always a preoccupation of fliers, found expression in the lines of one of the most beautiful airplanes ever built. This was the Supermarine S4 floatplane, which held the world speed record in 1925, achieving 226.6 mph (364.6 kmph), later models flying even faster.

Speed was subordinate to comfort in commercial machines, and the Dornier Dox was certainly roomy. The largest airplane in the world when it first flew in 1929, the Dox had six pulling and six pushing engines, two mounted on each nacelle. Unfortunately the giant —

Vickers Vimy
Cierva autogiro C6A
De Havilland Moth
Douglas DC2
Bell Model 47
Bell X-1
Mig - 15
Fairey Westland Rotodyne
SE 210 Caravelle
Tu-144
Bell X-15A

with a maximum flying weight of 123,460 lb (56,000 kg) and a wingspan of 157 feet 5 inches (48 meters) — was an expensive commercial failure.

Private fliers demanded a light, maneuverable airplane for use in the flying clubs that were opening in all parts of the world. The de Havilland Moth was one of the most successful of these small planes. In a Tiger Moth, Amy Johnson flew solo from England to Australia in $19\frac{1}{2}$ days in 1930.

The popularity of flying as a passenger service grew rapidly. In the DC series, the Douglas Co., in the U.S., produced one of the finest families of passenger aircraft. From the DC1 in 1933 to the DC3 Dakota — workhorse of World War II — the family gave remarkable international service. The Boeing 307 Stratoliner, in 1938, brought new comfort to high flight, enabling passengers to breathe easily in its pressurized cabin.

World War II brought a great surge in the development of aircraft speed, size and carrying power. By 1943, the Bell Model 47 helicopter was flying. The race to solve the problems of jet flight was fought neck-and-neck between the Allies and Germany, and, in 1944, Britain's Gloster Meteor became the first jet fighter to go into service.

The sound barrier, breached under controlled conditions by Chuck Yeager in 1947 flying a Bell X-1, marked another landmark in powered flight. Another aircraft from the same company, the Bell X-5, became the first airplane to fly with variable geometry wings, altering their angle in flight.

The de Havilland Comet, in 1949, brought commercial flying into the jet age. The graceful French Sud-aviation SE 210 Caravelle became the first passenger jet airliner to have its engines mounted on either side of the fuselage, a practice that became widespread later.

Russia's MiG 15, with its several variants, won a place as the world's most numerous fighter, and its qualities proved a serious shock to the fliers who fought it in the Korean War in 1951. Hawker Siddeley achieved one of the aims of military aircraft designers when they produced their V/STOL fighter, the Harrier. They made an aircraft that could take off and land without the aid of a runway, hover and fly at supersonic speeds, and so extend the role of a ground-support aircraft.

The next logical step in commercial aircraft was to build a supersonic passenger airliner. The Russian Tu-144 beat the Anglo-French Concorde into the air by a little over three months. The design of the two aircraft is extraordinarily similar. The urge to greater extremes of speed and height is still pursued. In 1967, the American X-15A rocket powered aircraft flew at 4,534 mph (7,247 kmph), and maneuvered above the Earth's effective atmosphere.

On 4 October 1957, the Russians led the way by taking a machine into space, orbiting Sputnik I around a wondering Earth.

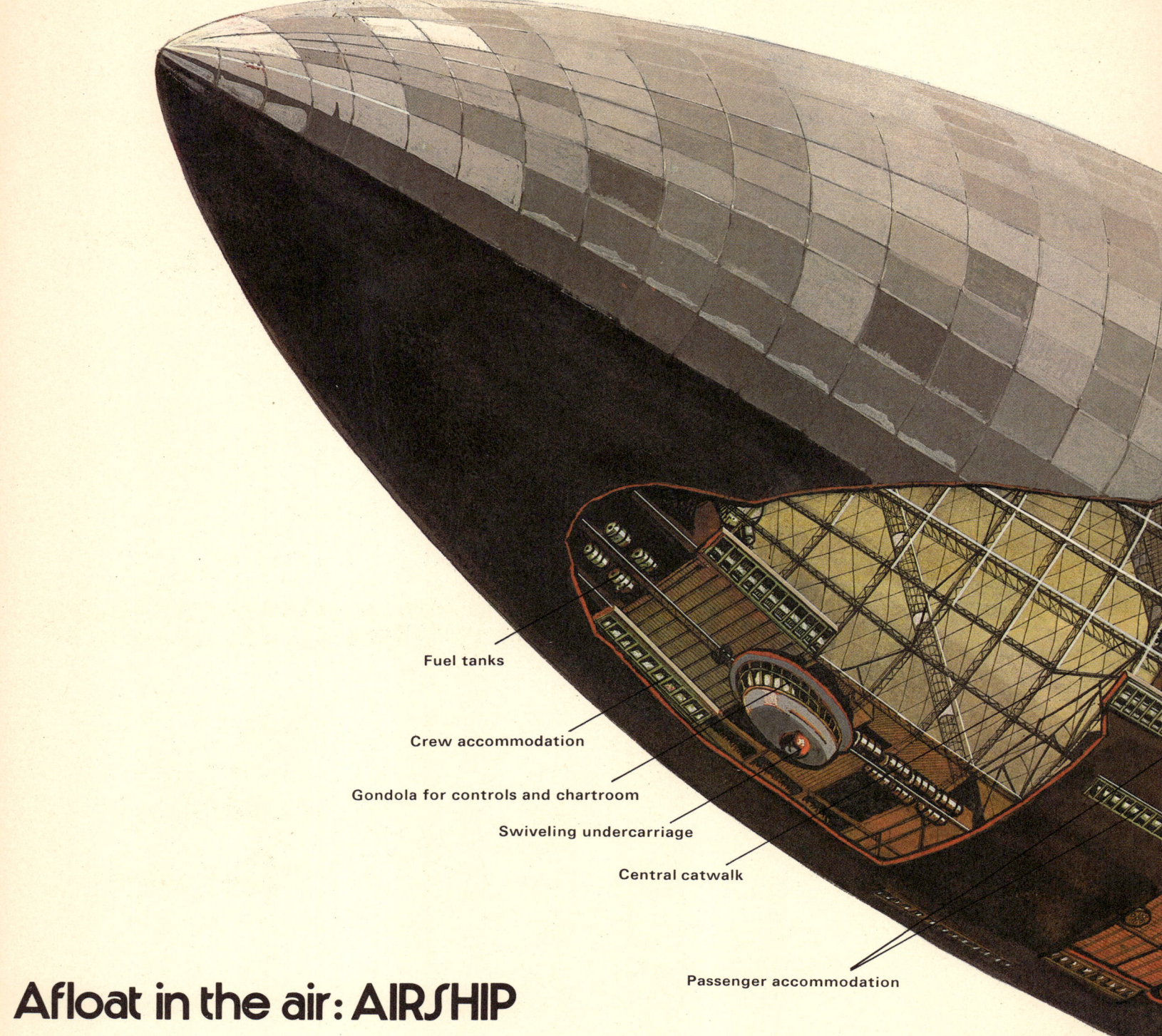

- Fuel tanks
- Crew accommodation
- Gondola for controls and chartroom
- Swiveling undercarriage
- Central catwalk
- Passenger accommodation

Afloat in the air: AIRSHIP

Ever since the first hot-air balloons of the Montgolfier brothers flew in 1783, men have dreamed of conquering the skies in large powerful airships. These lighter-than-air machines reached their peak of popularity in the 1930's but now, after an absence of nearly half a century, large airships are being planned as rivals to modern container ships.

Airships are steerable balloons pushed or pulled through the air by propellers driven by engines. The first, pioneered in France, used steam engines. They were called *dirigibles* (from a word meaning steerable) and were filled with the gas hydrogen. This is the lightest gas known, but it is also one of the most inflammable — it is currently used as a rocket fuel.

In 1899, the German Count Ferdinand von Zeppelin built a 470-foot long airship powered by two gasoline engines. Unlike earlier ships, which were merely flexible gas-bags with engines and a passenger compartment slung below, it was rigid. That is, it was built as a metal lattice frame with several hydrogen-filled bags anchored within it. By 1914, Zeppelin had established an airline which clocked up thousands of successful passenger-miles. During World War I, German Zeppelins bombed London.

After the war, Britain built passenger airships up to 650 feet long which carried people in luxurious comfort across the Atlantic Ocean. Then in the 1930s a series of disasters spelled doom for these clumsy lighter-than-air craft. First in 1930 the British R101 crashed in flames at Beauvais in France. Then the German *Hindenburg* exploded in the United States in 1937. The fierce heat of the burning hydrogen melted the ship's metal like butter within seconds, and many people were killed. The United States continued to use *blimps*, as they called them, during World War II, but the commercial passenger airship was finished.

In 1975, British engineers tested a *flying saucer* shaped airship. There are again plans for monster cigar-shaped ships for carrying heavy freight between industrial centers, without the need for large port or airport facilities.

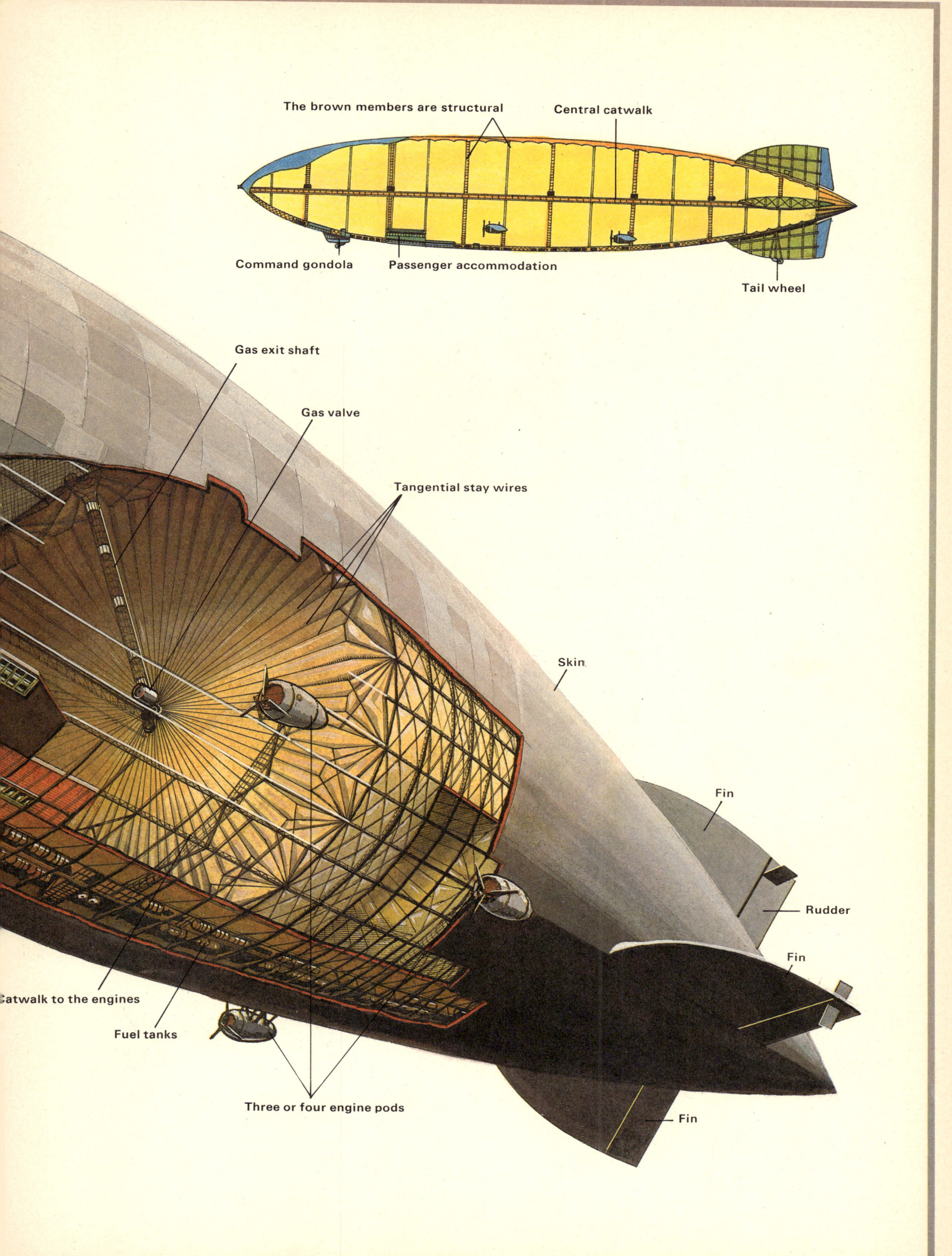

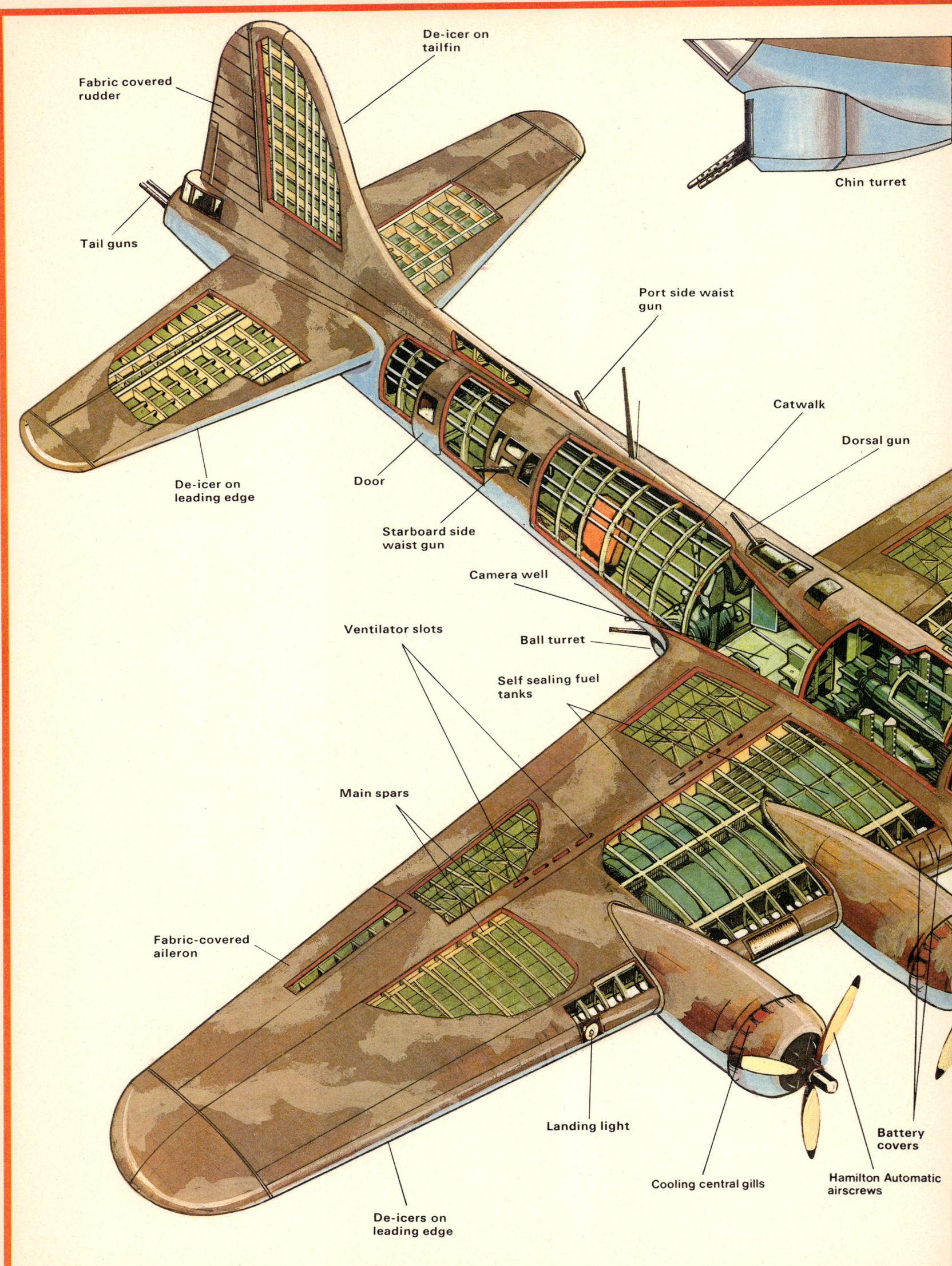

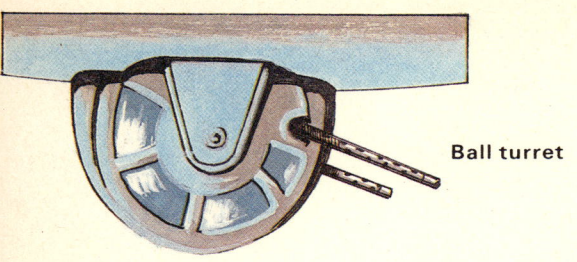
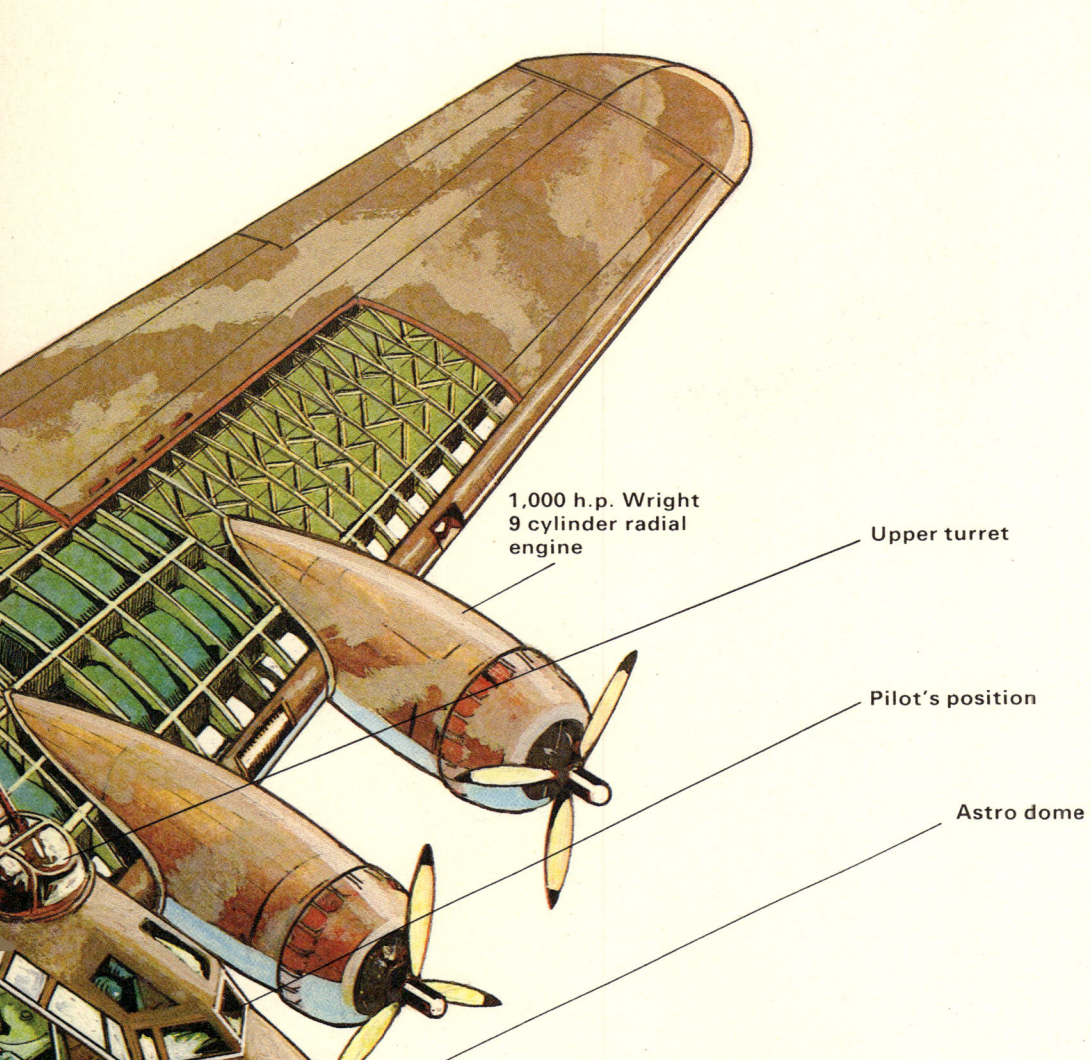

Airborne destroyer: FLYING FORTRESS

The Boeing B-17 *Flying Fortress* represented the peak of mass-produced bomber development in World War II. Its speed and armament meant that it could outfly and outgun many fighters. Equipped with four 1000-horsepower engines and an oxygen system it could fly at high altitudes, beyond the range of most anti-aircraft guns. By 1944, the United States had built 12,700 B-17s.

From about May 1943 bombers of World War II flew in huge formations of up to a thousand aircraft to carry a sufficient payload for saturation bombing. Even then, they made easy targets and needed strong fighter protection or had to fly at night, which led to problems with navigation and bomb-aiming. The *Flying Fortress* was designed to "take care of itself". Upper and lower twin gun turrets were manned by machine gunners. The radio operator and navigator fired other guns if necessary. And two more twin-gun turrets in the tail and *chin* (under the nose) of the aircraft could be aimed and fired by gunners lying in those uncomfortable positions.

De-icers on the leading edges of the wings, tail-plane and tail-fin prevented ice building up on them in the freezing temperatures at high altitudes. Turbochargers gave fuel economy and an extra 200 horsepower from each engine for take-off. Gasoline was carried in fuel tanks in the wings, which were self-sealing if pierced by bullets.

The successor to the *Flying Fortress*, the B-29 *Superfortress*, carried the atom bombs which were dropped on Hiroshima and Nagasaki in 1945 to bring World War II to a close.

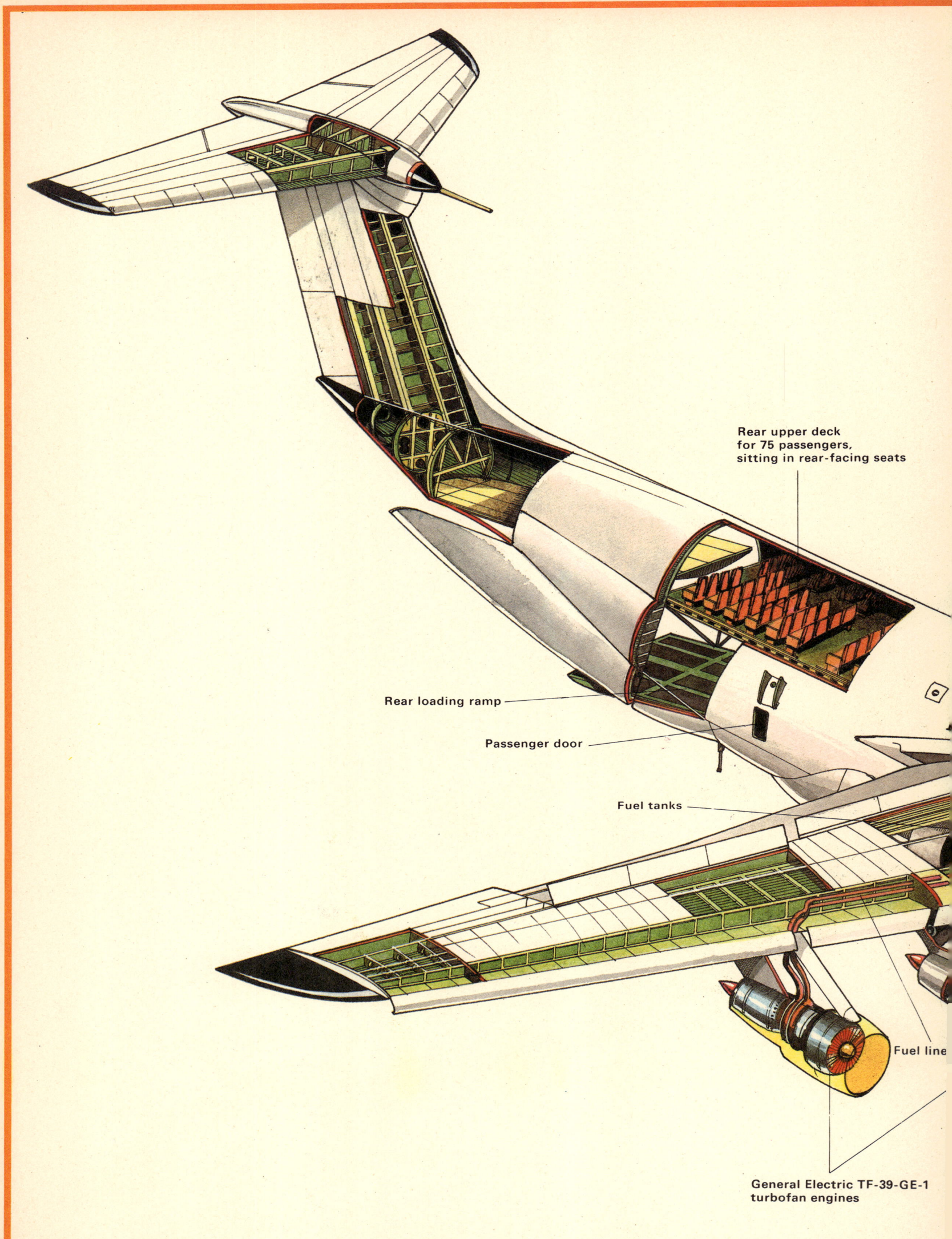

Giant cargo plane: GALAXY

In the 1960s, the United States Army urgently needed a new, large cargo plane. The aircraft industry responded by producing the enormous Lockheed C-5A, known as the Galaxy. From drawing board to the first operational aircraft delivered, took a mere forty months of intensive work. The airplane they gave to the Army was a monster that could carry two M48 tanks — which weigh 99,000lb (45,000kg) each — or three Chinook helicopters, across transatlantic distances. Military airstrips are not always large or elaborate, so the Galaxy's take-off distance of 7,000 feet (2,134 meters) and its landing distance of 2,230 feet (680 meters) are conveniently short, considering the great size of the airplane.

The Galaxy needs only a five-man crew, a pilot, co-pilot, flight engineer, navigator and a cargomaster. This last crew member is obviously essential when considering the variety and size of the cargo the aircraft carries. The rear upper deck will accommodate 75 soldiers, and a further 270 can be seated on the lower deck if necessary. Air-conditioning of the pressurized interior of the airplane gives the passengers a comfortable ride, but the lower deck is usually kept clear for cargo. As many as 16 three-quarter ton trucks may be secured in this area. Above the cargo deck, there is accommodation for another 15 crewmen who may act as relief for the flight crew. The aircraft is most vulnerable while on the ground, so the designers have given it a hinged visor-type of nose and a large ramp at the rear, enabling the cargomaster to arrange loading and unloading quickly and, if needed, simultaneously. The Galaxy can also be used in making airdrops.

Lockheed C-5A Galaxy

Fuselage: aluminum alloy and titanium alloy 230ft 7in. (70.29 meters) long
Engines: 4 G.E. TF 39-GE-1 turbofans
Fuel: 20 fuel tanks; total usable capacity of 49,000 US gall (185,450 liters), total usable oil capacity of 36.4 US gall (138 liters)
Wingspan: 222ft 8in. (67.88 meters)
Length overall: 247ft 10in. (75.54 meters)
Height overall: 65ft 1in. (19.85 meters)
Basic operating weight: 337,937lb (153,085kg)
Maximum take-off weight: 769,000lb (348,400kg)
Maximum landing weight: 635,850lb (288,040kg)
Payload: 220,967lb (100,100kg)
Range with a load of 112,600lb (51,000kg) is 6,529 miles (10,505km)
Speed at maximum take-off weight is 472mph (760kmph)
Speed at 25,000ft (7,630 meters) is 571mph (919kmph)
Average cruising speed: 581mph (934kmph)
Flight ceiling: 34,000ft (10,360 meters)

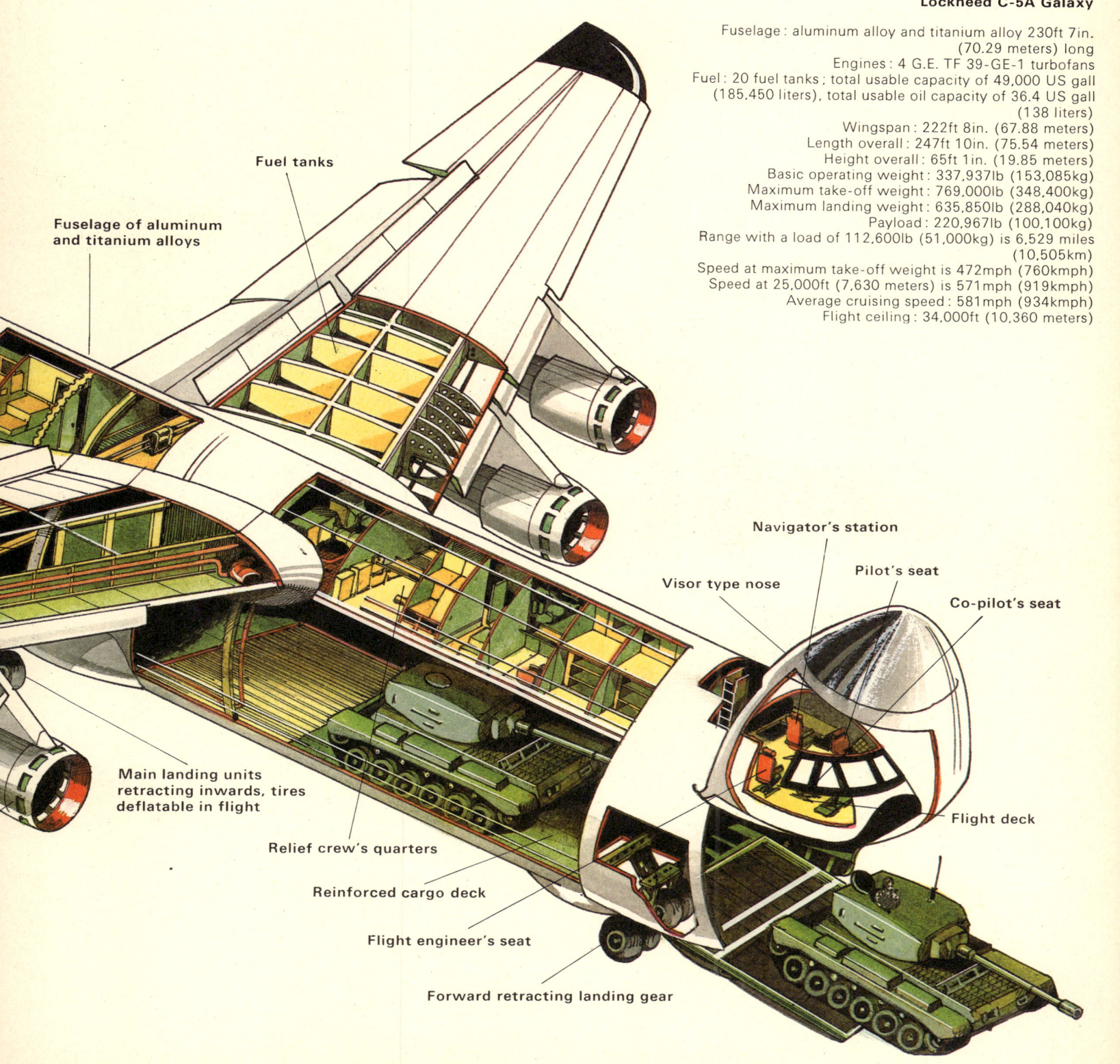

Transatlantic giant: JUMBO JET

The Boeing 747 is the world's largest passenger aircraft in service, hence its popular name Jumbo. Its excellent safety record and the ease of its operation makes it popular with crews and air companies throughout the world. There was some initial opposition to the aircraft caused by loading problems, and the lengths of runway required for this enormous aircraft. It needs a minimum of 6,750 feet (2,057 meters) of runway at its maximum landing weight.

Since the first 747 left the ground in 1969, several versions have been developed to meet various commercial needs. Some are passenger planes and others are strictly freight carriers, but its design is such that conversions from one use to another are fairly simple to arrange. Despite its great size and the sophistication of its design, it requires only the normal three-man crew that smaller airliners carry, although the large numbers of passengers demand a much larger cabin staff. A Jumbo can carry a maximum of 364 passengers.

Boeing 747

Wingspan: 195ft 8in. (59.64 meters)
Length overall: 231ft 4in. (70.51 meters)
Length of fuselage: 225ft 2in. (68.63 meters)
Height overall: 63ft 5in. (19.38 meters)
Tail plane span: 72ft 9in. (22.17 meters)
Passenger plane weight: 370,816lb (167,979kg)
Passenger plane payload: 155,684lb (70,617kg)
Maximum level speed at 30,000ft (9,144 meters): 608mph (978kmph)
Cruising ceiling: 45,000ft (13,716 meters)
Range with maximum fuel: 7,090 miles (11,410km)
Radius of turning circle on land: 75ft (22.9 meters)

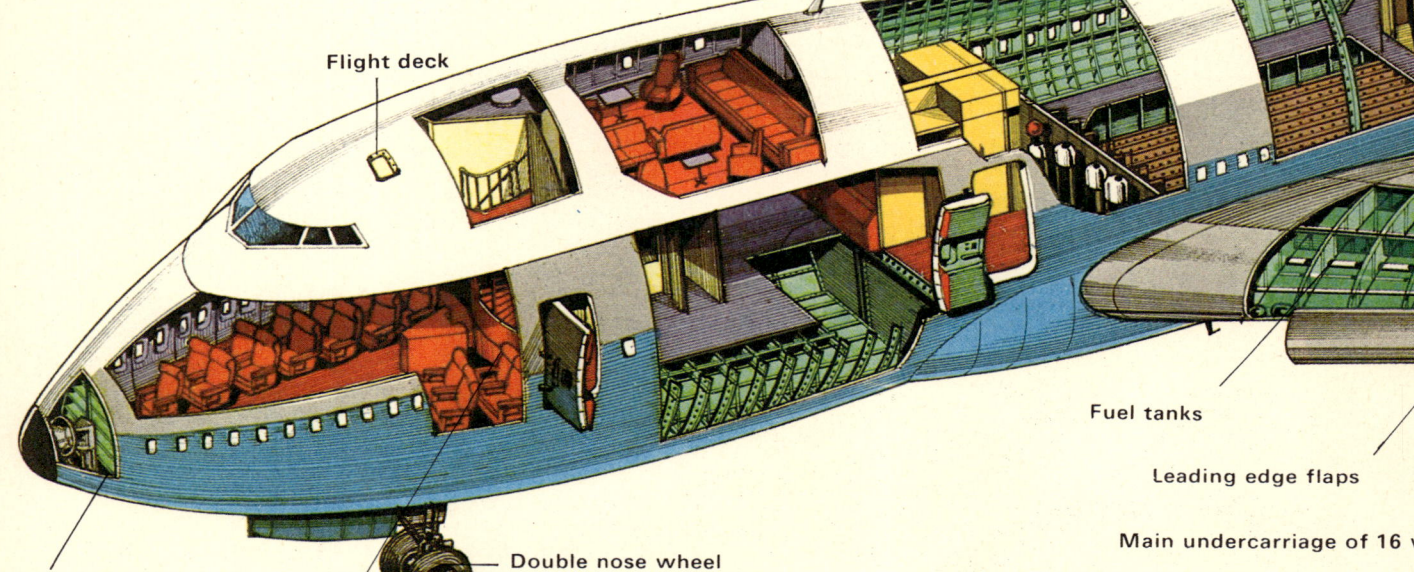

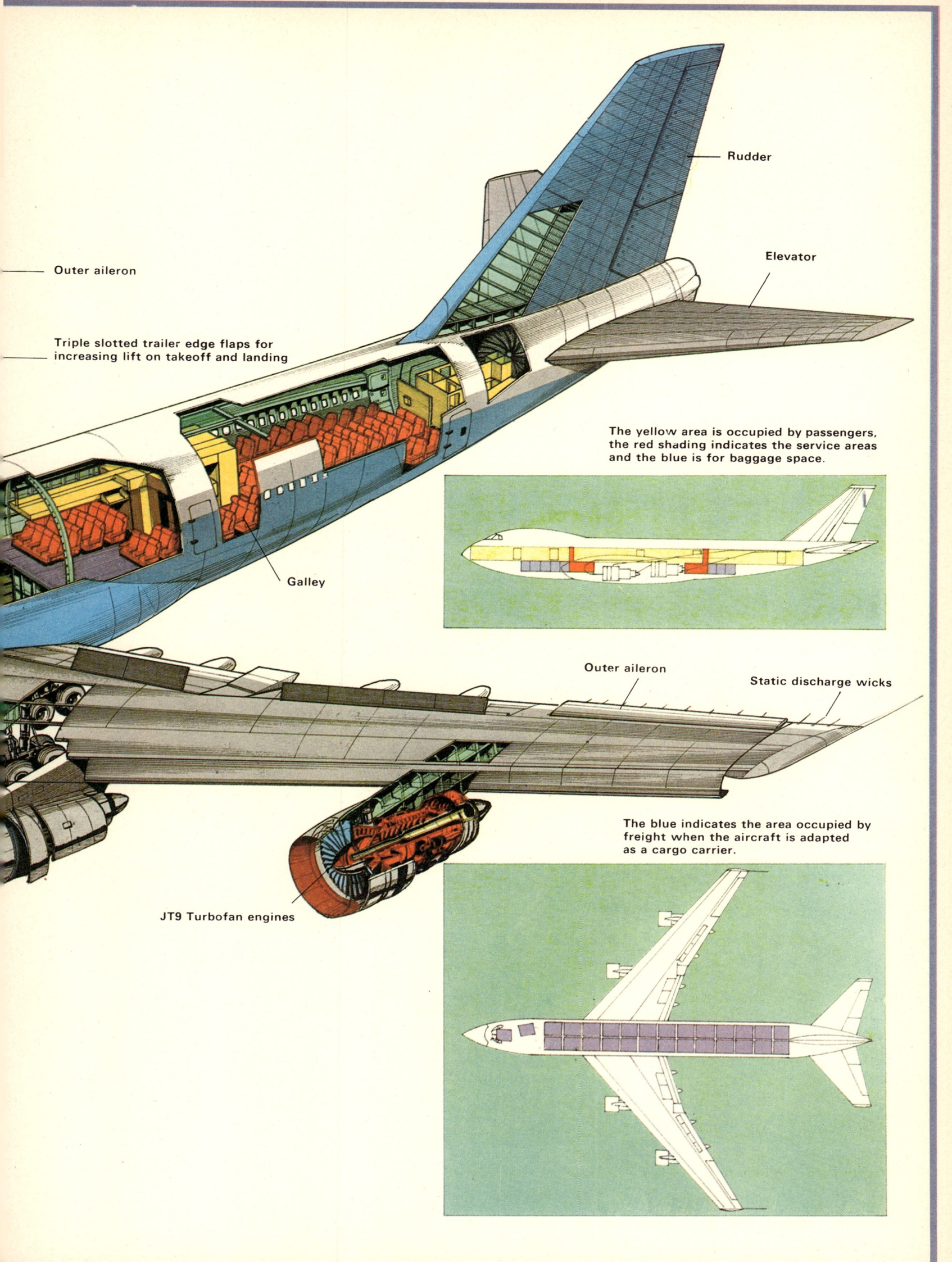

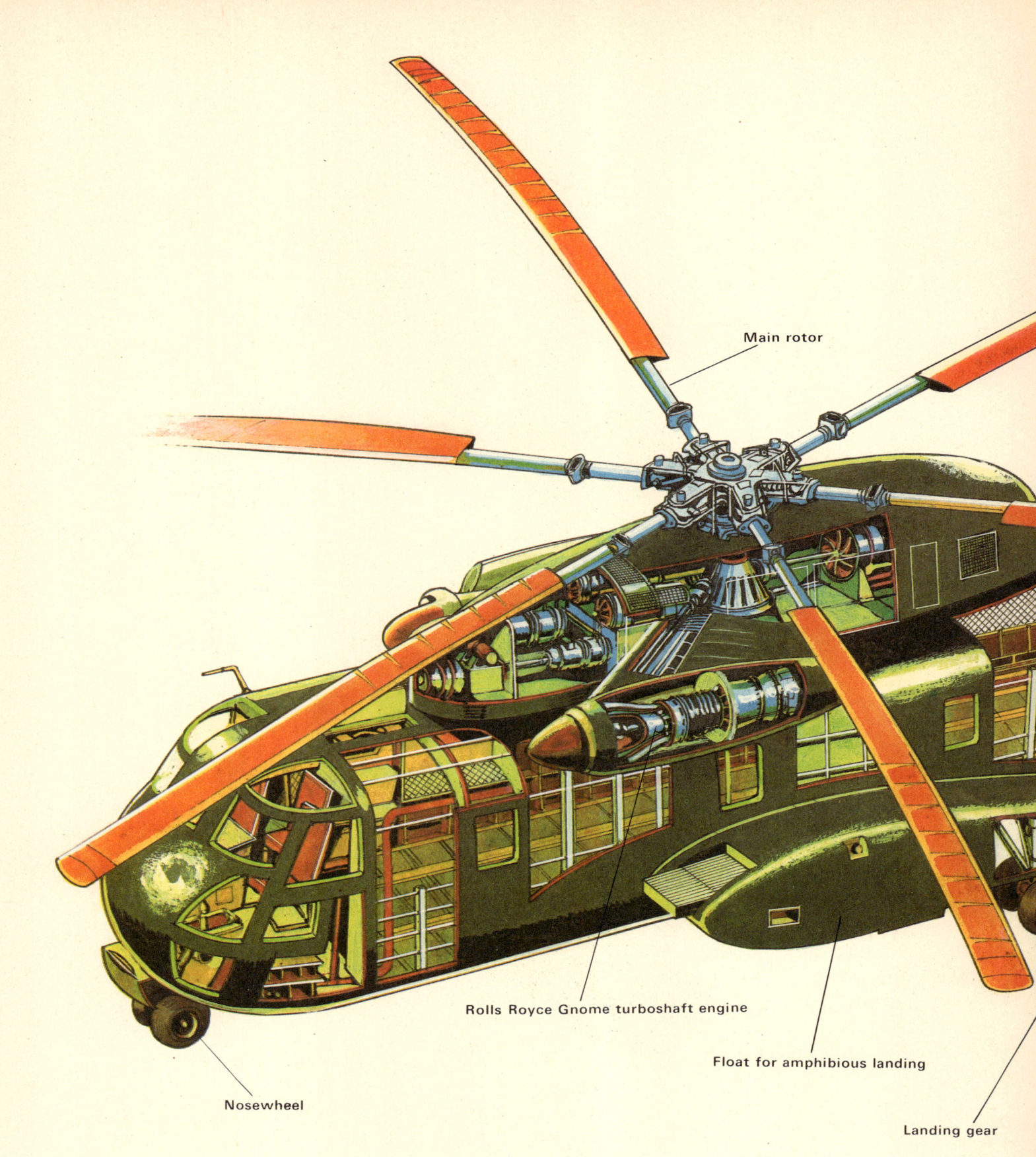

Sikovski Sea Stallion

Fuselage length: 67ft 2in. (20.47 meters)
Height: 24ft 11in. (7.6 meters)
Engines: two 3.925shp G.E. T64-GE-413 turboshaft
Rotor diameter: 72ft 3in. (22.02 meters)
Maximum take-off weight: 41,000lb (19,050kg)
Maximum speed: 196mph (315kmph)
Hovering ceiling: 13,400ft (4,080 meters)
Range: 257 miles (413km)

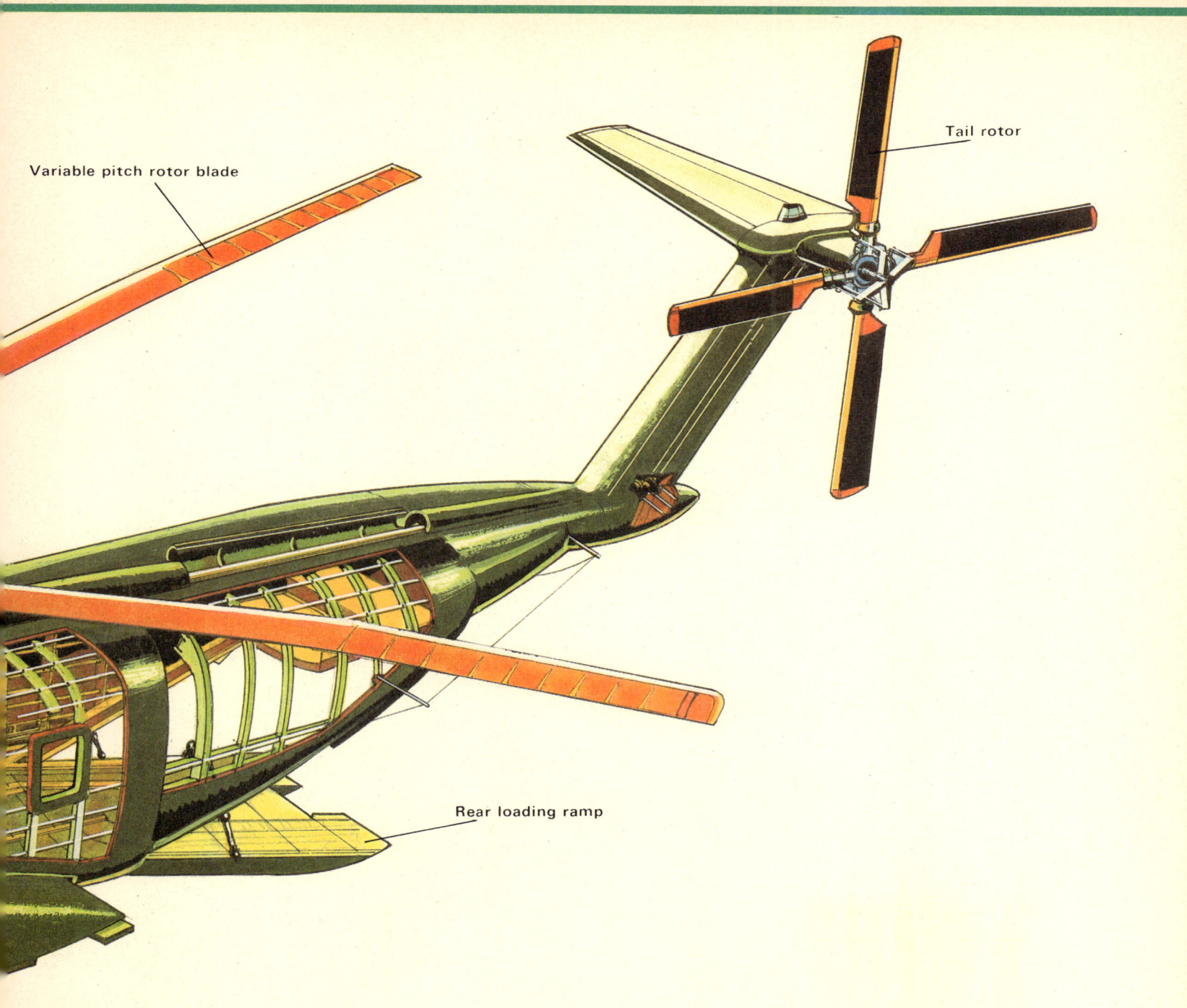

Whirling wings: HELICOPTER

One of the first practical helicopters was the VS-300, designed and built by Russian-American Igor Sikorsky in 1939, whose company later became a world leader in helicopters. One of its biggest technical advances was achieved in the S-61, originally built as an anti-submarine aircraft in 1959.

This drawing shows a version of the Sikovski Sea Stallion, which carries equipment for detecting and attacking submarines. Another version is the Commando Army helicopter, which does not have amphibious landing gear, and yet another is the 28-seat S-61 N commercial transport which is one of the types used to ferry men and supplies to North Sea oil rigs in all weathers. All helicopters have the basic disadvantage that they have to lift and drive themselves along by a revolving rotor—a set of rotating wings—which means that in forward flight they are less efficient than airplanes. On the other hand, they can do things that are impossible for airplanes; they can operate from any reasonably flat platform, and can hover. In hunting for submarines, they can hover just above the sea, listening with various sensing devices—mainly acoustic sonobuoys—and finally drop homing torpedoes at exactly the right spot. Army helicopters carry men and supplies, and can dodge about behind trees or other cover while "killing" tanks with guided missiles. Crane helicopters can place spires on churches or air-conditioning systems on tall office buildings, and airline helicopters serve the centers of many cities. Gradually the helicopter has overcome the problems of high cost, high insurance risk, short fatigue life and a noisy and vibrating ride to become a thoroughly well-developed and attractive vehicle. Much of the credit rests with modern turbine engines, which have constantly become quieter and more efficient. Over the 21 years since 1955, helicopter speeds have risen from 65 to almost 200mph (104 to 322kmph) and payload from 9 per cent of the empty weight to over 80 per cent.

Harrier Jump Jet

One 21,500lb (9,740kg) thrust Rolls-Royce Pegasus 103 vectored-thrust turbofan, with four nozzles able to rotate over angle of 98° from rearwards direction
Span: 25ft 3in. (7.70 meters) extends to 29ft 8in. (9.04 meters) by adding bolt-on *ferry tips*
Length: single-seater 45ft 6in. (13.87 meters); two-seater 55ft 9in. (17 meters)
Height: single-seater 11ft 3in. (3.43 meters); two-seater 13ft 8in. (4.17m)
Empty weight: (typical) 12,000lb (5,443kg)
Gross weight: over 25,000lb (11,339kg)
Maximum speed: over 737mph (1,186kmph)
Mach number in dive 1.3
Ceiling: over 50,000ft (15,240 meters)
Range with one in-flight refueling: over 3,455 miles (5,560km)

- Laser range finder
- Bird proof windshield
- Cockpit canopy
- Steerable ducts
- Air intake for engine
- First stage fan blades
- Rolls Royce Pegasus vectored thrust turbo fan engine
- Nose wheel
- 30mm Aden gun pod
- 30mm Aden gun
- Outer pylon
- Main gearbox
- Ammunition box
- Wingtip landing gear

Various positions of the Harrier's engines

Take-off

Landing

In flight

Hover power: HARRIER JUMP JET

By 1960, every builder of combat aircraft was sketching projects for fighters and attack aircraft that could leap out of any small area of flat ground, lifted by jet thrust. In that year Hawker Aircraft in Britain first flew a small research aircraft called the P1127, and today this has been developed into the only jet V/STOL (vertical/short take-off and landing) aircraft in operational use.

Compared with the P1127, the Harrier is much more powerful. No two parts are common to the two designs, but both share a supreme simplicity in having only one engine. The Pegasus engine was designed to discharge compressed air from left and right front nozzles and hot gas from left and right rear nozzles, and all four nozzles are coupled together and driven by a pneumatic motor to point downwards for lift, forwards for aerial braking and rearwards for ordinary flight. Compressed air from the engine, of very high temperature, is taken through small pipes to nozzles at the wing tips, nose and tail where special valves can allow it to blast in any chosen direction. By this means the pilot can control the Harrier even when hovering, when ordinary control surfaces would be useless.

The first Harriers went into service with RAF squadrons in April 1969, and about 200 were supplied to the RAF and U.S. Marine Corps, some being two-seat trainers. Though basically used for ground attack and reconnaissance, they have been found to be amazingly tricky customers in the fighter role, carrying a 30mm gun and missiles and using their swiveling nozzles to make incredible maneuvers. Production in 1975 was beginning on the Harrier FST1 for the maritime role aboard various ships, with new radar and weapons. Hawker and the U.S. company, McDonnell Douglas, have also designed a more advanced aircraft, the AVI 16A, which might go into use in the 1980s. Only gradually have armed services realized the potential of combat jets that do not need an airfield.

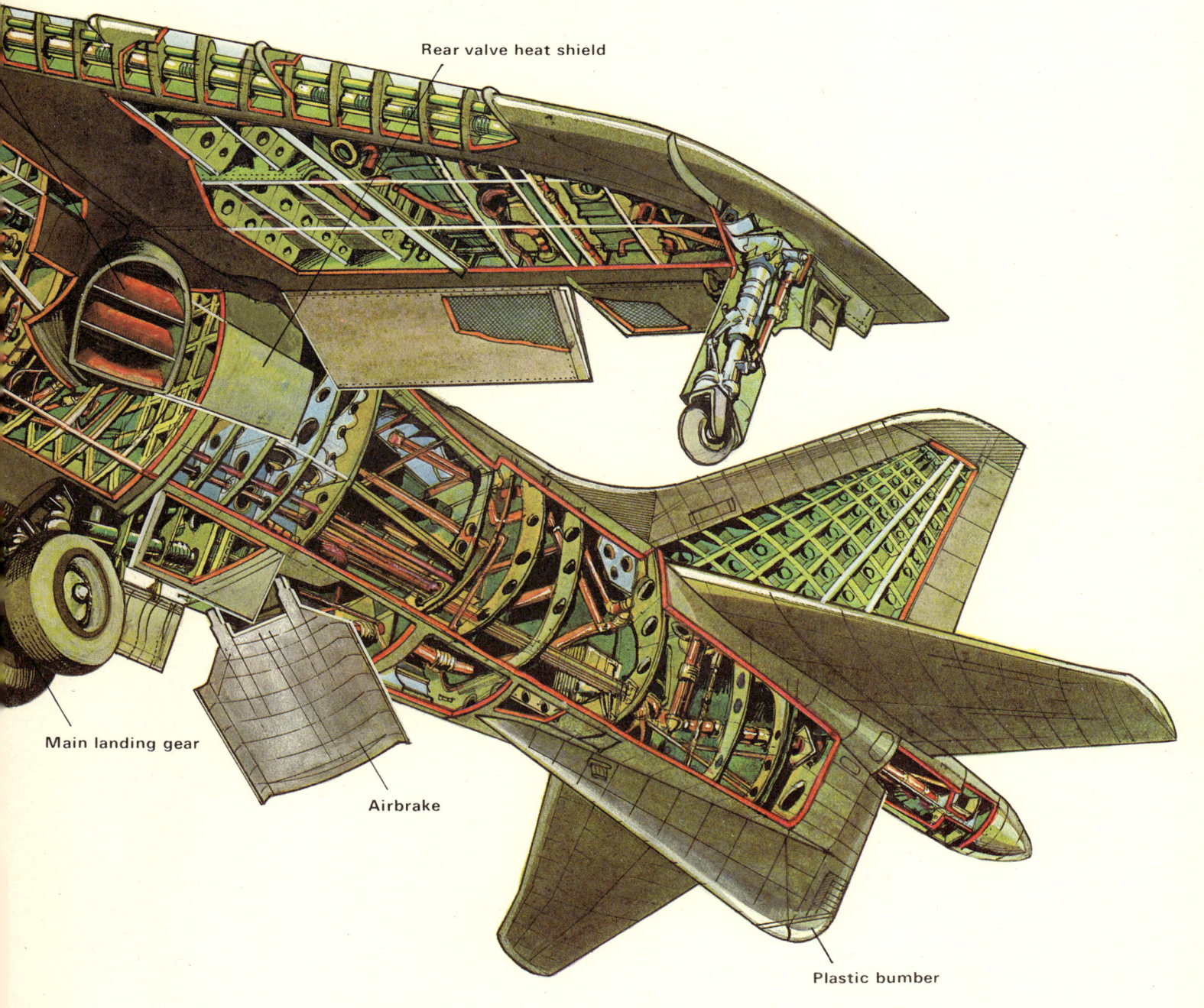
Rear valve heat shield
Main landing gear
Airbrake
Plastic bumber

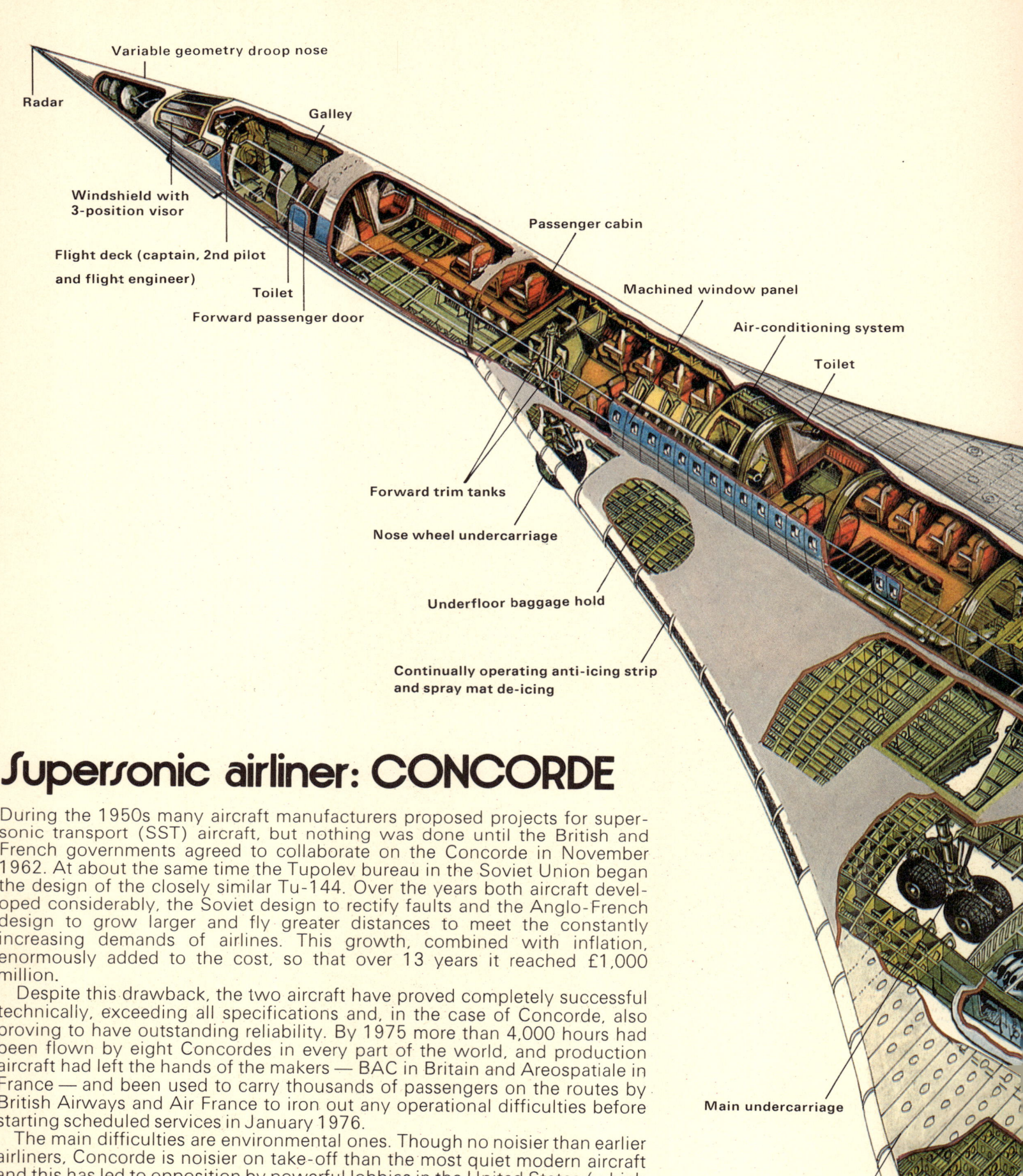

Supersonic airliner: CONCORDE

During the 1950s many aircraft manufacturers proposed projects for supersonic transport (SST) aircraft, but nothing was done until the British and French governments agreed to collaborate on the Concorde in November 1962. At about the same time the Tupolev bureau in the Soviet Union began the design of the closely similar Tu-144. Over the years both aircraft developed considerably, the Soviet design to rectify faults and the Anglo-French design to grow larger and fly greater distances to meet the constantly increasing demands of airlines. This growth, combined with inflation, enormously added to the cost, so that over 13 years it reached £1,000 million.

Despite this drawback, the two aircraft have proved completely successful technically, exceeding all specifications and, in the case of Concorde, also proving to have outstanding reliability. By 1975 more than 4,000 hours had been flown by eight Concordes in every part of the world, and production aircraft had left the hands of the makers — BAC in Britain and Areospatiale in France — and been used to carry thousands of passengers on the routes by British Airways and Air France to iron out any operational difficulties before starting scheduled services in January 1976.

The main difficulties are environmental ones. Though no noisier than earlier airliners, Concorde is noisier on take-off than the most quiet modern aircraft and this has led to opposition by powerful lobbies in the United States (which has no aircraft able to compete). It has also been suggested that high-flying aircraft might in some way harm the atmosphere. Another problem is caused by the sonic boom — a sharp double bang — on the ground directly below an aircraft flying faster than sound, so supersonic routes have to avoid populated areas. The ideal routes are transoceanic, London to New York taking exactly half the seven hours needed by the fastest subsonic jets. In 1974 a Concorde left Boston as a Boeing 747 left Paris, flew to Paris, spent an hour on the ground and then landed at Boston before the 747.

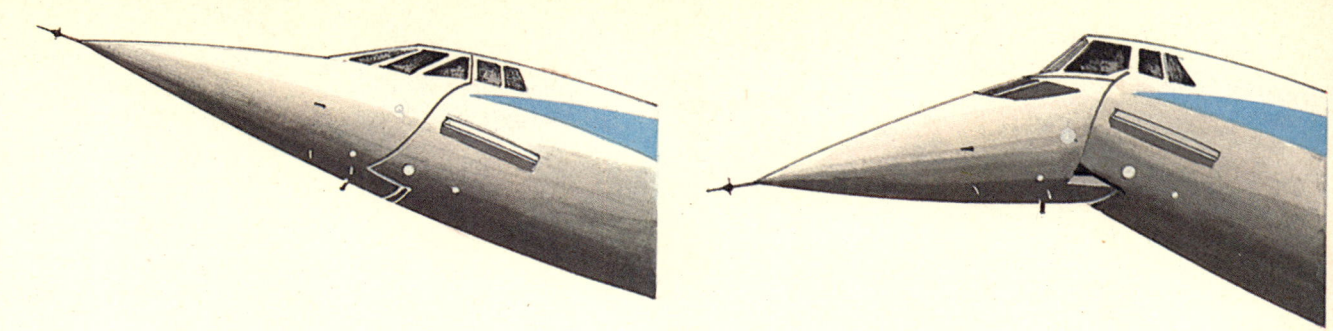

The variable geometry droop nose of Concorde allows the pilot to make a visual landing and to maneuver visually on the ground.

Concorde SST

Length: 202ft 1in. (61.66 meters)
Wingspan: 83ft 10in. (25.56 meters)
Gross weight: 400,000lb (181,200kg)
Engines: 4 Rolls Royce (Bristol) SNECMA Olympus 593 turbojets of 38,050lb (17,260kg) thrust each
Seats: 100-128
Speed: Mach 2.05 (1,353mph)
Range: 4,000 miles (6,400km)

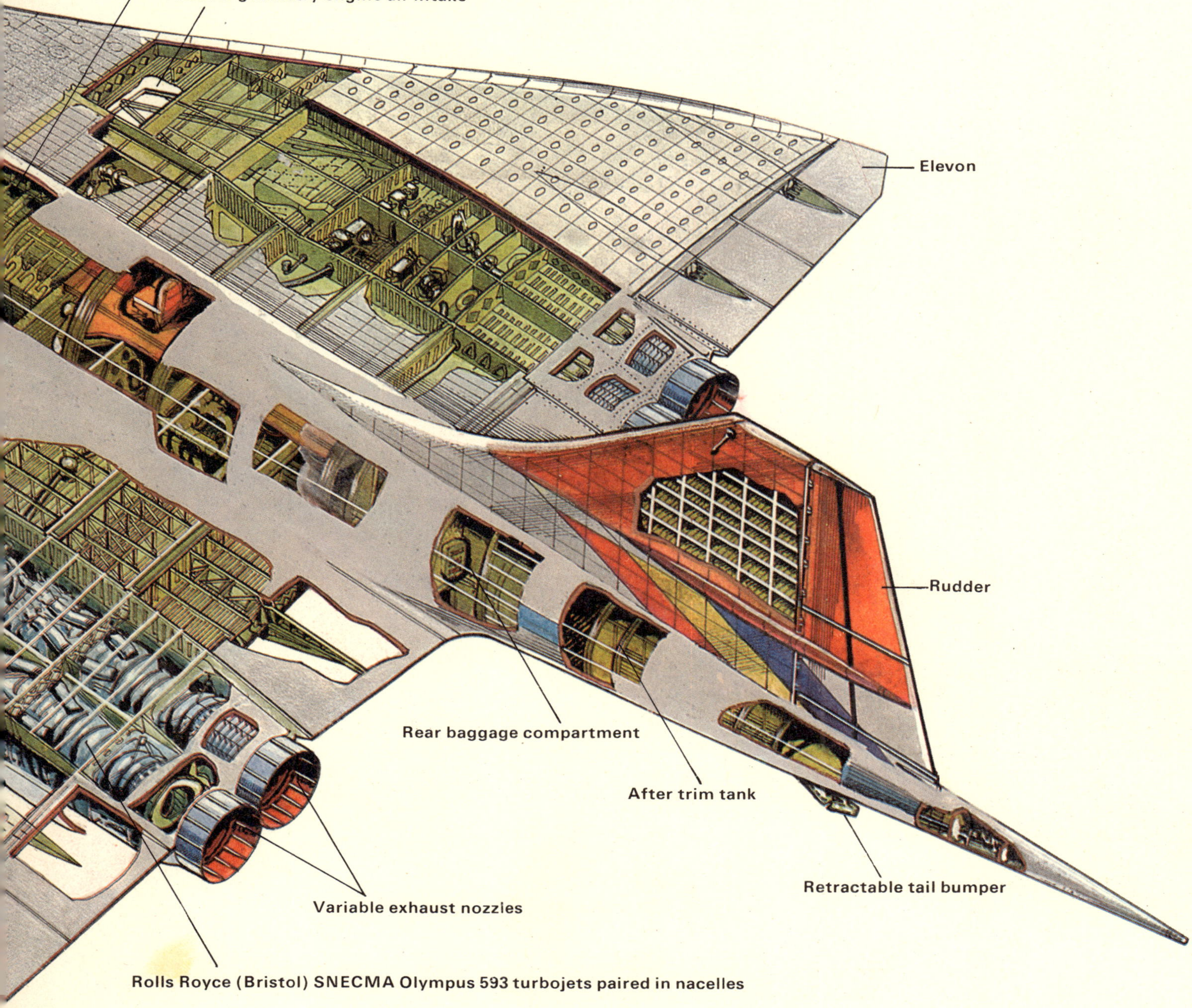

- Pressurized floor above wheel bay
- Variable geometry engine air intake
- Elevon
- Rudder
- Rear baggage compartment
- After trim tank
- Retractable tail bumper
- Variable exhaust nozzles
- Rolls Royce (Bristol) SNECMA Olympus 593 turbojets paired in nacelles

21

Man into space: APOLLO

Project Apollo achieved one of man's longest-held dreams when two American astronauts landed on the Moon in 1969. Although the method they used to fly to the Moon was complicated and dangerous, five more landings were made. One other mission was abandoned after an explosion in the spacecraft, but the astronauts returned safely to Earth.

The Apollo spacecraft was launched into space by the mighty Saturn V booster, the whole rocket standing 363 feet (110.6 meters) high. The spacecraft itself was in three sections. The conical command module was the control center of the spacecraft and housed the three astronauts for most of the mission. Behind it was attached the cylindrical service module, which contained the main power and oxygen supplies and the main engine. The spidery lunar module made the descent to the Moon with two astronauts aboard, while the other astronaut remained in the command module in orbit round the Moon. Of the great Moon rocket that left the Earth, only the small command module returned. To bring back the booster and other modules would have required a huge amount of fuel and a much bigger rocket. It is easier, in spaceflight, to throw something away when it no longer has a use rather than use up precious fuel in returning it to Earth.

During the landings, the astronauts spent many hours exploring the Moon's surface, setting up scientific instruments and taking rock samples. They found the Moon to be a totally dead world and, although the samples showed it to be about the same age as the Earth, it was probably never part of the Earth but another body that formed separately and was long ago *captured* by the Earth. The Apollo astronauts used a Lunar orbit rendezvous method to fly to the Moon. The Apollo spacecraft was fired on its way to the Moon after an orbit of of the Earth. During the journey, the command module linked up with the lunar module. Then the spacecraft went into orbit around the Moon and the two astronauts entered the lunar module, separated from the command module and descended to the Moon's surface.

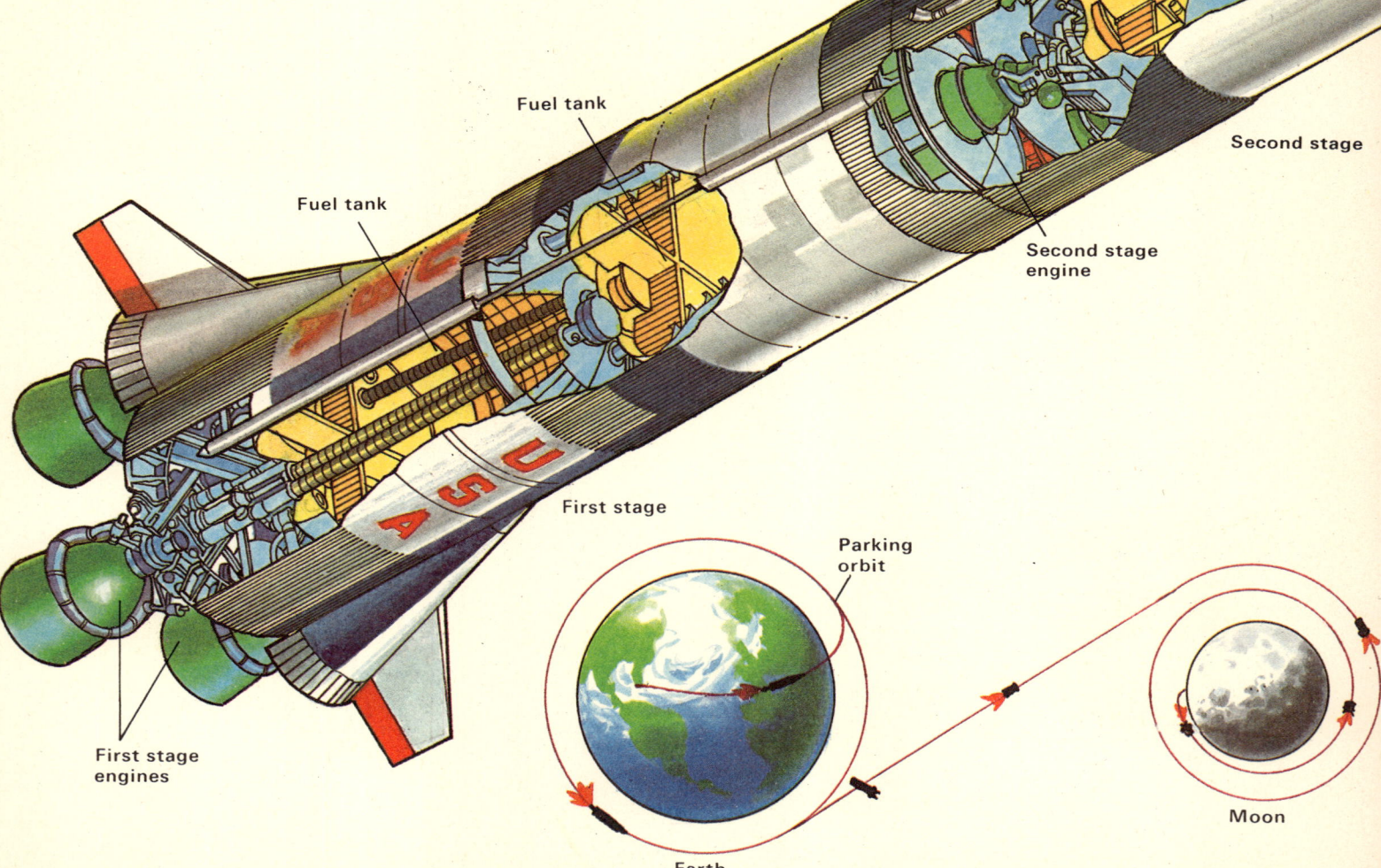

Launch
escape
tower
Command
module
Service
module
Lunar
module
Third stage
Fuel tank
Third stage
engine

Lunar
module
Docking
hatch
Command module
Tunnel

Fuel tank
Thrusters
Docking
probe
Parachute
Astronauts
Engine
nozzle
Radio antenna
Thrusters

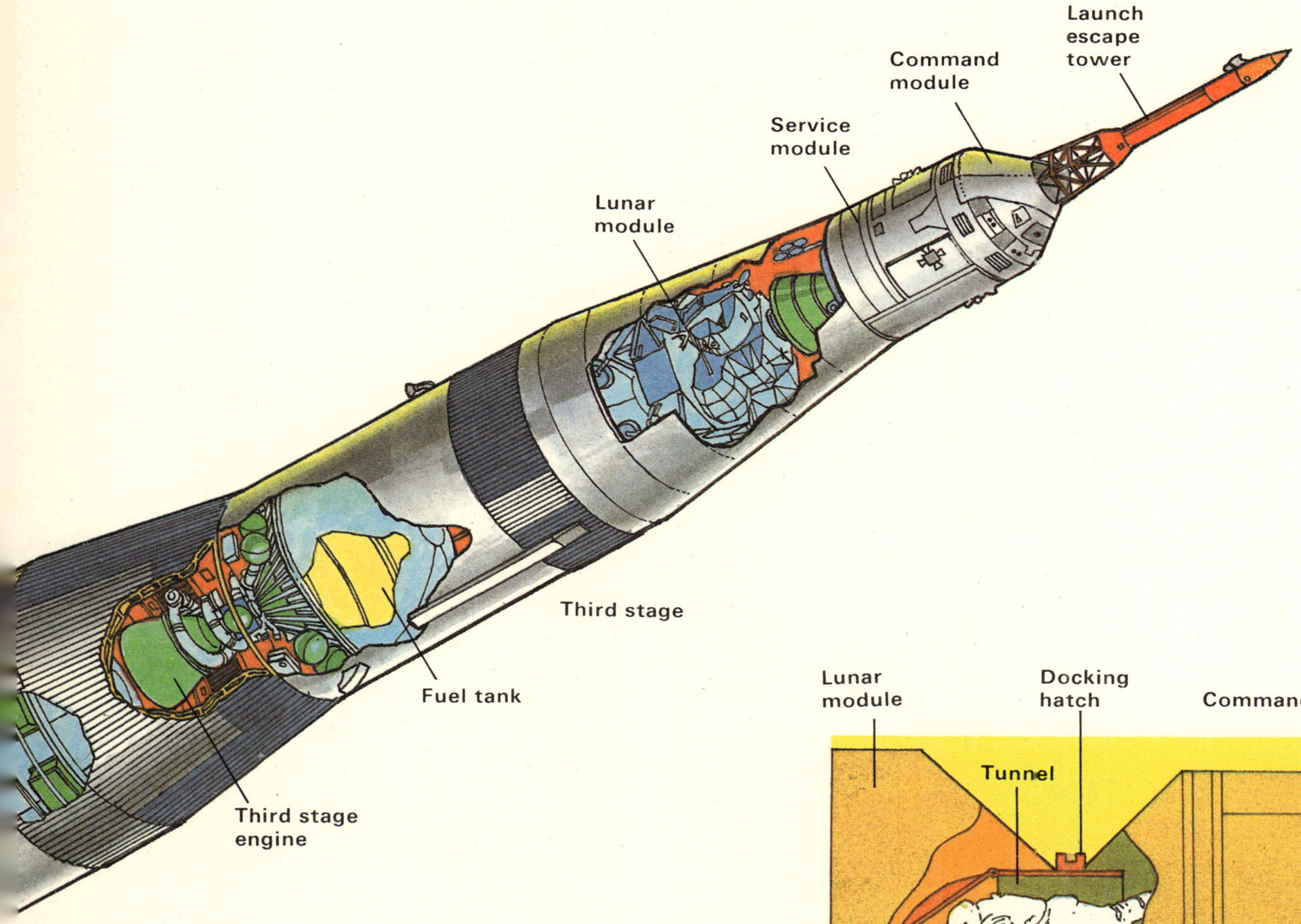

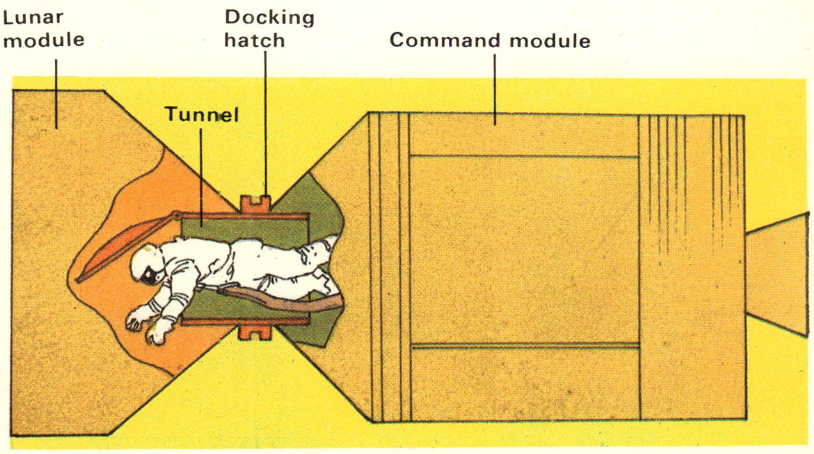

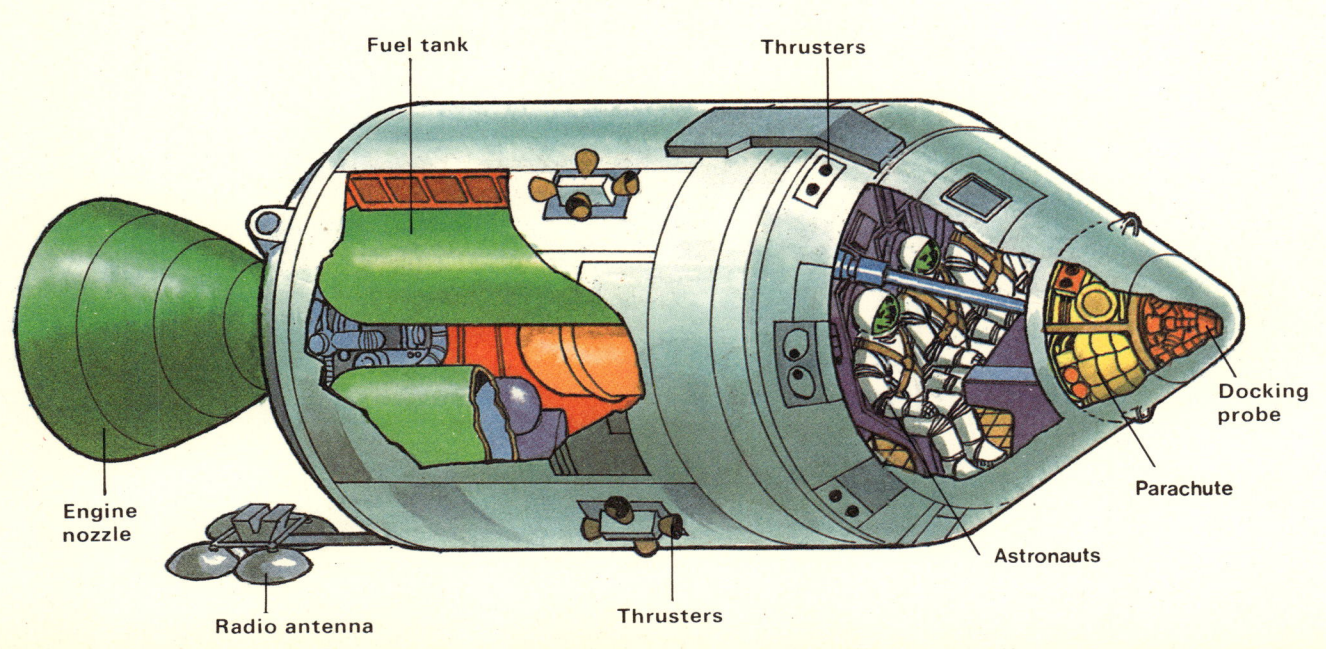

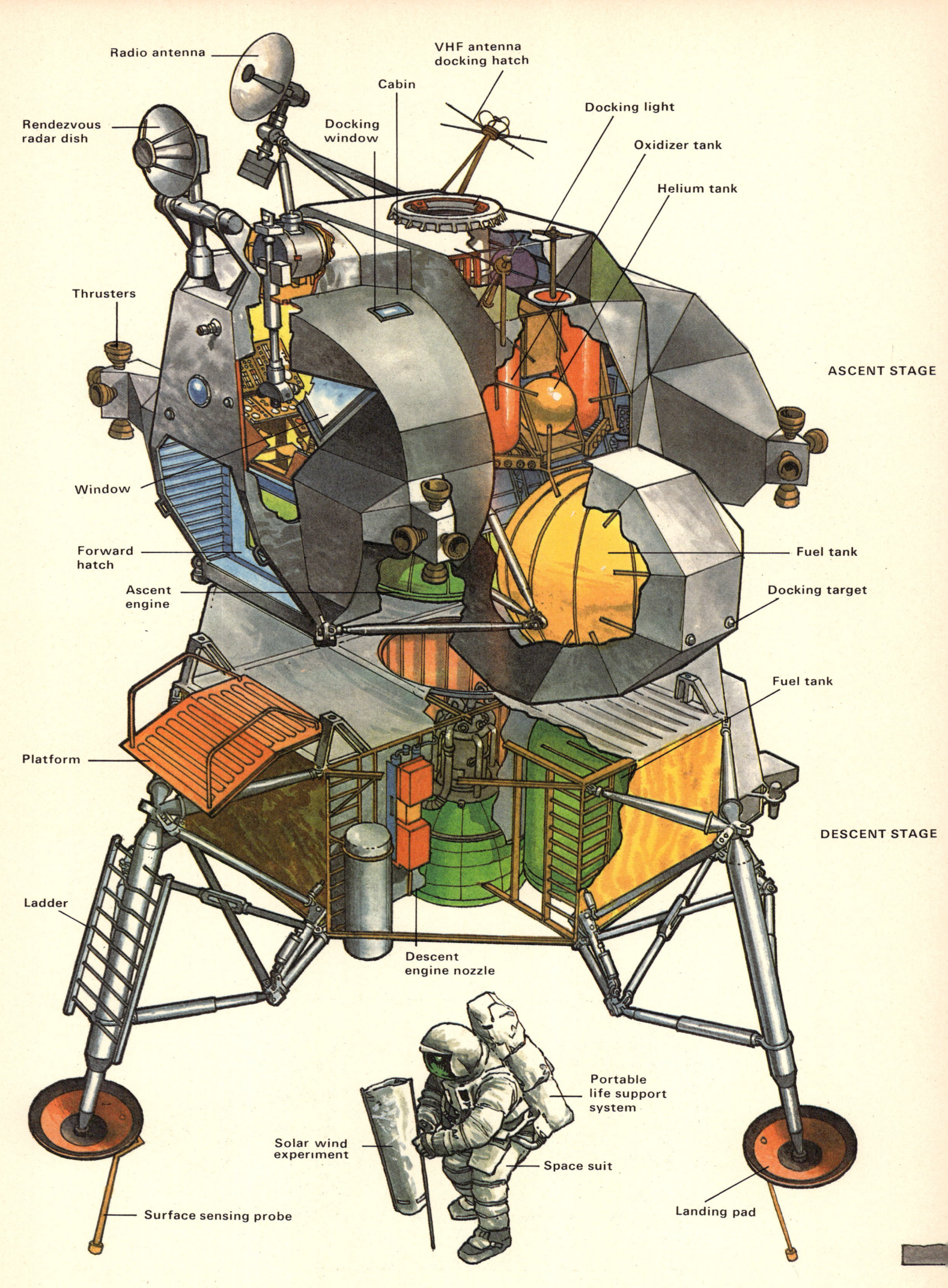

Moon landing: LUNAR MODULE

The lunar module of the Apollo spacecraft had none of the smooth, sleek lines that we normally expect of spacecraft, but this is because it was designed to work only in space and did not need to be streamlined. It consisted of two parts: an ascent stage containing a cabin for the two astronauts, and a descent stage beneath it.

While in orbit around the Moon, the astronauts entered the lunar module from the command module through the hatch at the top. They then separated from the command module and descended to the Moon by firing the engine in the descent stage. Probes beneath the four feet told the pilot when he was about to land. The two men then put on their spacesuits and left the module through the main hatch at the side and climbed down the ladder to the surface. They unloaded equipment from storage bays in the descent stage and set about their exploration of the Moon. The later Apollo missions carried a Moon car to drive over the surface, but the first astronauts had to walk. They set up instruments to detect such things as Moonquakes, magnetic fields and particles from the Sun, and radio the information back to Earth; the instruments were powered by nuclear generators so that they continued to work long after the missions were over. The astronauts also took many rock samples and carried them back to Earth. Because there is no air on the Moon, the astronauts talked to each other by radio as they worked. Their spacesuits gave them oxygen to breathe and protected them from the heat and glare of the Sun. Although the suits were heavy and bulky, the spacemen worked easily because the Moon's low gravity gave them only a sixth of their Earth weight.

Their work over, the astronauts left the Moon by firing the engine in the ascent stage. The descent stage was left on the Moon and served as a launching pad. The lunar module rejoined the command module in Moon orbit and the two astronauts transferred back to it. Then the lunar module was abandoned and the main engine in the service module fired to send the spacecraft back to Earth. The service module was also abandoned just before re-entry into the Earth's atmosphere, leaving the command module to splash down into the ocean beneath its parachutes.

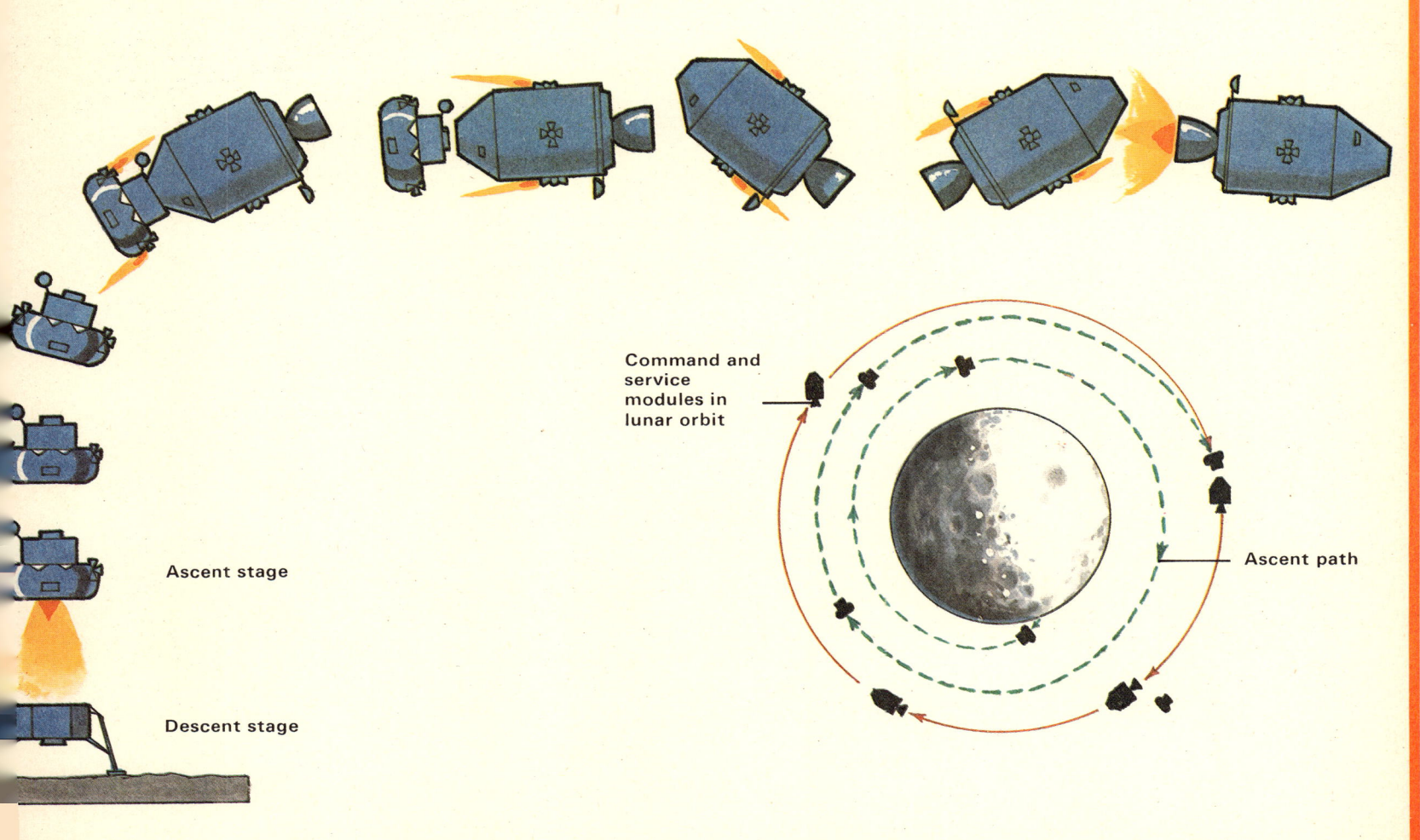

Ascent stage

Descent stage

Command and service modules in lunar orbit

Ascent path

Covered wagon
Sleigh
Steam car
Early gas engined car
Modern sports car
Subway train
Semitrailer truck
Transcontinental diesel train

ON THE GROUND

While fire was probably Man's greatest discovery, his greatest invention was, without any doubt at all, the wheel. Without this fundamental achievement travel on land would have remained forever painfully slow, dependent on river boats, sledges and horses only. The applications of the wheel are so wide that it would be hard to name an aspect of technology in which the wheel did not have an important function.

No one knows who invented the wheel, but its first appearance was in Sumer in Mesopotamia over 5,000 years ago, and its use spread rapidly across the world. There were, however, peoples who remained unaware of its uses. The Indians of North America had no knowledge of the wheel until Europeans introduced covered wagons to the great plains of their hunting grounds. Even when the Indians had adopted the European's horses, they continued to drag behind them a V-shaped frame as a carrier of goods. The sledge continues in use in parts of the world today, not only running on ice but on grass too.

The wagon, chariot and carriage all developed quickly. The solid wheel was abandoned for a lighter, stronger spoked wheel. Even by the first century BC, Scandinavian craftsmen had perfected a hardwood bearing which reduced friction on the axle to give a faster and more reliable ride. This they managed with small hardwood rollers inserted between the outside of the axle and the inside of the hub.

Comfort and efficiency improved as the craft of the wheelwright developed. The elegant carriages of the eighteenth century were sprung by suspending the bodywork from leather straps so that the movement over ill-made roads was changed from bone-shaking jolts to a more bearable sway. Regular changes of teams of horses allowed good speeds over the poor roads. In nineteenth century England, a record run from London to Brighton and back — about 108 miles (173.8 kilometers) — was completed in an average speed of 13.79 mph (22.1 kmph). This achievement required eight teams of horses and 14 changes.

John Loudon McAdam's development of a new road surface in 1815 transformed the traveler's experience from an exhausting, bruising trial to a smooth and delightful experience, wherever he was fortunate enough to meet the new surface.

Other means of driving a vehicle along were a major preoccupation from the second half of the eighteenth century. Indeed, as early as 1668, Ferdinand Verbeist, a Belgian Jesuit missionary to the Court of the Emperor of China, made and ran a model steam automobile. But it was not until the Frenchman, Nicholas Cugnot, made a three-wheeled steam tractor that there was a vehicle strong enough to carry people, and which could run without animal power.

A hundred years of experiment and improvement followed. In 1885, Karl Frederich Benz ran his first gasoline-driven car between eight and ten miles per hour (12.87 – 16.09 kmph) at Mannheim. The next 90 years saw the introduction of new power units and tremendous technological advances. By 1970, Gary Gabelich was briefly topping the 650 mph (1,046 kmph) mark in his Blue Flame, a rocket powered land vehicle. But the family car had been established by Benz's invention and Henry Ford's manufacturing methods, when he mass produced the Model T Ford. Between 1908 and 1927, he sold 15,007,033 of them to a world hungry for personal transport.

Man-powered road vehicles had been popular among daring spirits as early as the 1640s, when

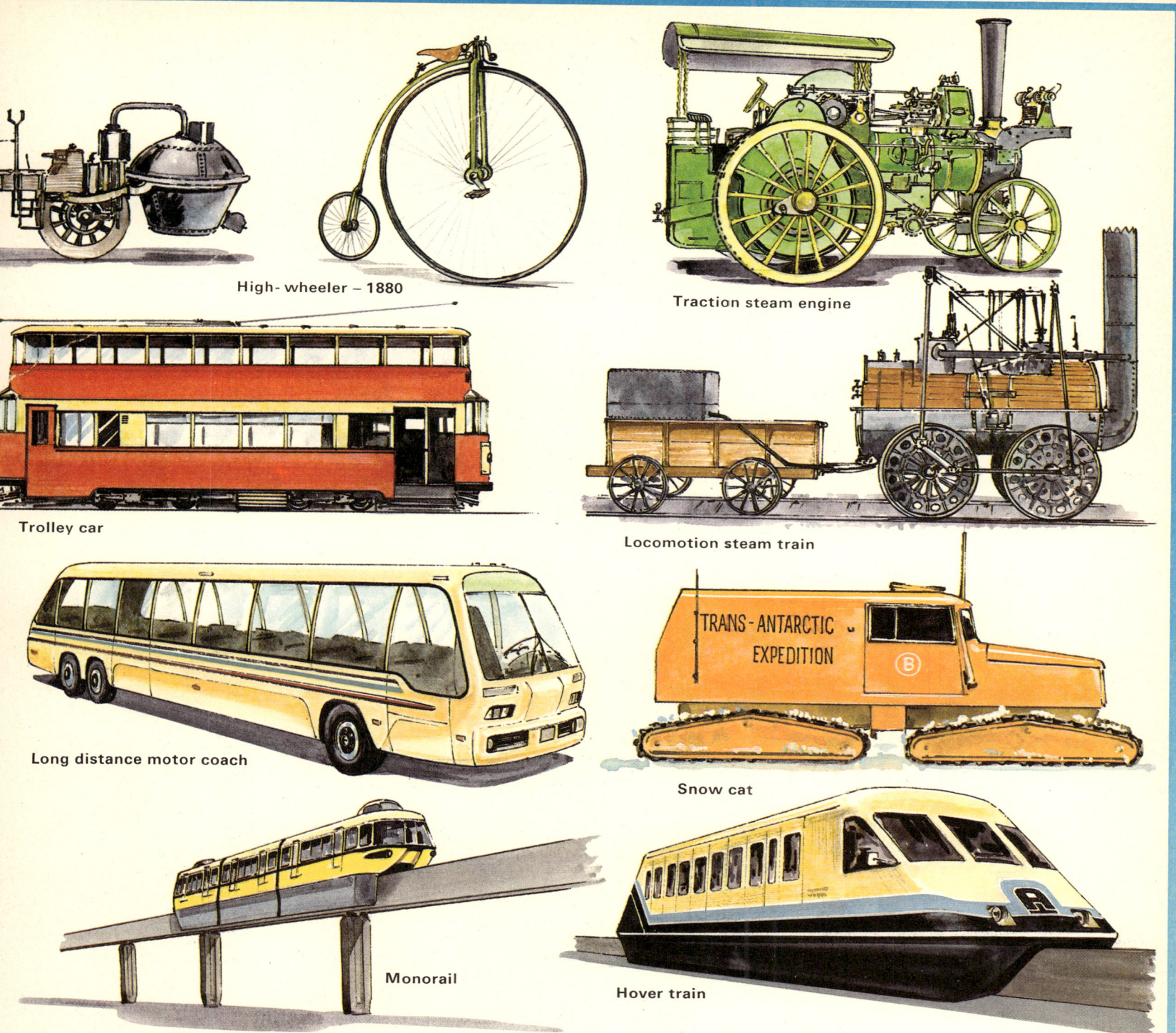

High-wheeler – 1880
Traction steam engine
Trolley car
Locomotion steam train
Long distance motor coach
Snow cat
Monorail
Hover train

velocipedes were scooted down hill by young men who did not care to think of anything so far beneath their notice as brakes. By 1839, the bicycle, cranked by pedals, was a reality. Soon bloomer-clad ladies braved the catcalls as they pedaled through dusty lanes.

The urgent need for mass transport for the people of industrial Europe and America encouraged the growth of public interurban companies. The first of these ran between Eastbourne and Meads in England in 1903. It was an idea that soon spread far from sleepy Sussex. The culmination of the interurban idea is the Greyhound motor coach, running 3,240 miles (5,213 km) between San Francisco and Miami in the U.S. They achieve a remarkable average speed of 39.59 mph (63.7 kmph) for their enormous run, and accomplish it in stylish comfort.

The efficient transport of materials is as important to modern living standards as the comfortable and safe movement of people. The steam traction engine puffed its way into the place left by the passenger steam carriage, which had been made redundant by the gasoline combustion engined car. The traction engine was a reliable and remarkable versatile vehicle. It could be used as a stationary power source in agriculture or as a strong workhorse for hauling loads.

Soon the gasoline and diesel truck ousted the steam giants. Designers and manufacturers were restricted in the size of their vehicles only by the capacity of roads to take them. The largest of these mighty trucks, the Rotinoff tractor Super Atlantic, produces as much as 400 bhp. Specialized transport was designed for especially demanding types of terrain. For swamps, vehicles with enormous, balloon-like tires kept them near the surface of the ground, while for Arctic transport the Snowcat proved stronger and more reliable than the time-tested dog team and sledge.

The railroad had, since the mid-nineteenth century, proved its unassailable effectiveness in transporting enormous loads and thousands of passengers in safety over great distances. *Locomotion* the well-named locomotive built by George Stevenson in 1825 was one of a 'line' of steam engines, but it became one of a *stable* of locomotives on the first public railroad. Steam engines helped to unite new nations and to administer the colonies in Africa and Asia. Steam provided energy for railroads throughout the nineteenth and the early twentieth centuries, but then the search for a cleaner and more economical source of power led to the application of diesel and electric powered engines to locomotives. More recently, experimental locomotives have been designed to use the hovercraft principle on rails. This is a line of development that is far from its end.

For safety's sake: FAMILY CAR

With the acceptance of the motor car as the most convenient means of personal and family travel, the numbers of cars on the road has increased dramatically. This has brought a consequent increase in the number of accidents, so safety features in motor car design are assuming greater and greater importance. A car that incorporates many new features is the Volvo 240.

Volvo cars have always been designed with safety in mind. They introduced the shatterproof windshield in 1944, and have included seat belts as standard equipment since 1959.

Some of the new safety features on the Volvo 240 are dual circuit brakes — the car has two braking systems, one operating on both front wheels and the left rear wheel, and the other operating on both front wheels and the right rear wheel. If one system should fail, say through a hydraulic brake failure, the other system will still work, providing 80 percent of full braking effect. In addition, disc brakes are fitted to all four wheels.

To avoid damage to the car at low speeds, the car's energy absorbing bumpers will withstand an impact of up to 3 mph (5 kmph). In the event of a head-on collision at speed, the steering column collapses, avoiding injury to the driver's chest. The engine housing and trunk are designed to absorb the energy from a collision at up to 43 mph (70 kmph) without the passenger area sustaining any damage, and the doors have been strengthened to withstand impacts from the side.

It has been known for some time that driver fatigue and discomfort have an adverse effect on his performance and, therefore, safety. The designers of this car produced orthopedically designed seats, with adjustable lumbar support to suit drivers of any size. Further adjustment is also provided in the tilt adjustment and in the height of the seat.

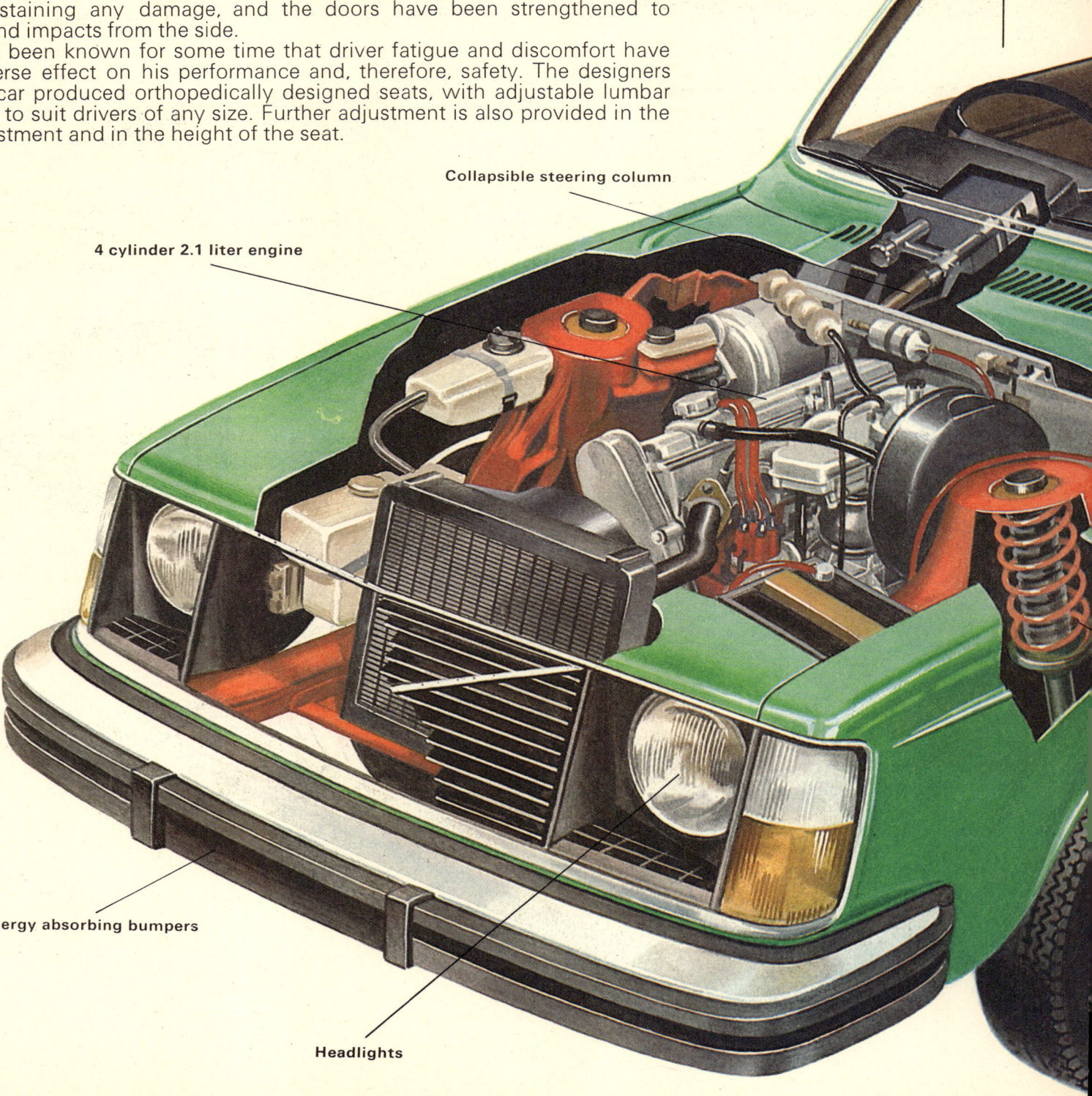

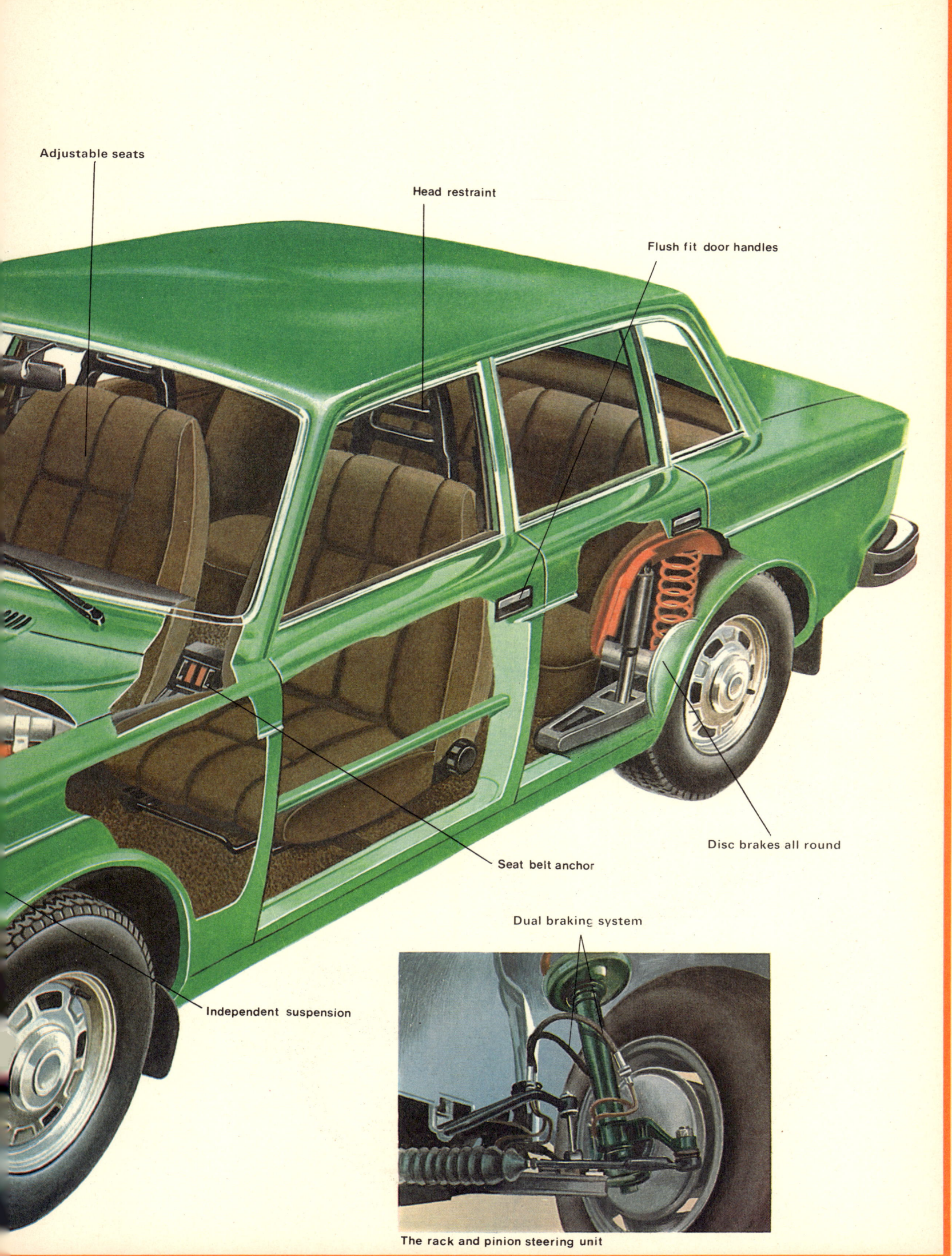

Super bus: GREYHOUND

The first horse-drawn omnibuses clip-clopped over the cobbled streets of Paris and London in the 1820s. The invention of the gasoline-engined car—by Karl Benz and Gottlieb Daimler in 1885—and then the diesel engine by Rudolf Diesel in 1897, provided tireless, adaptable motor power. Motor buses were able to rival railroads, and became part of the way of life of town and country dwellers throughout the world. Today, the peak of development of this form of transport is the high-speed luxury coach, typified by the Greyhound buses which each year carry thousands of passengers in comfort along the expressways of the United States.

A modern Greyhound bus is nearly 40 feet long and 8 feet wide (12.2 meters x 2.44 meters). Weighing more than 12 tons, it has a 285-horsepower diesel engine which allows it to carry up to 50 passengers at 70 miles an hour (112 kmph). Passengers enter by a door at the front, and most of the cushioned seats recline for extra comfort or sleeping. Hand baggage can be carried in racks over the seats, and a large storage space under the floor holds up to 300 cubic feet of suitcases and other bulky baggage. Central heating and air conditioning keep the temperature inside the bus in the low 70s F even if it is below freezing at or sweltering over 100 degrees F outside. There is a lavatory at the rear of long-distance intercity coaches.

For safety, the bus rides on eight wheels. In addition to the two steerable wheels on the front axle and four on the independently-sprung rear driving axle, there is an extra pair on the rearmost unpowered trailing axle. This gives better braking power and more contact with the road than earlier four-wheel designs. Shatterproof windows, a roof escape hatch, two fire extinguishers and a first-aid kit complete the safety features.

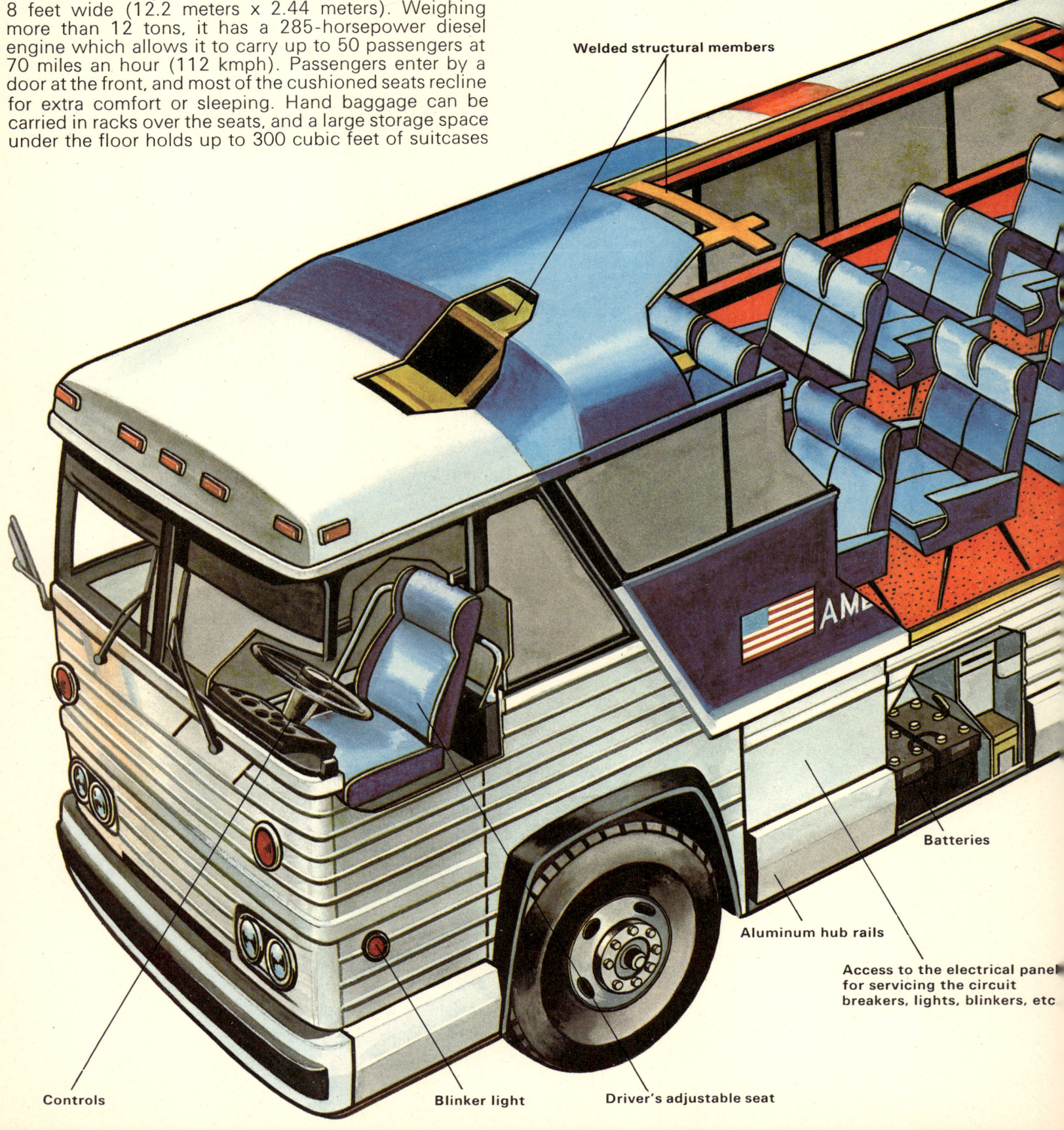

MC8 Crusader Greyhound

Length : 39ft 11in. (12.17 meters)
Width : 8ft (2.43 meters)
Height : 10ft 10in. (3.3 meters)
Underfloor luggage capacity : 300cu ft (8.5cu meters)
Turning radius : 48ft 4in. (14.72 meters)
Weight : 26,760lb (12,122kg)

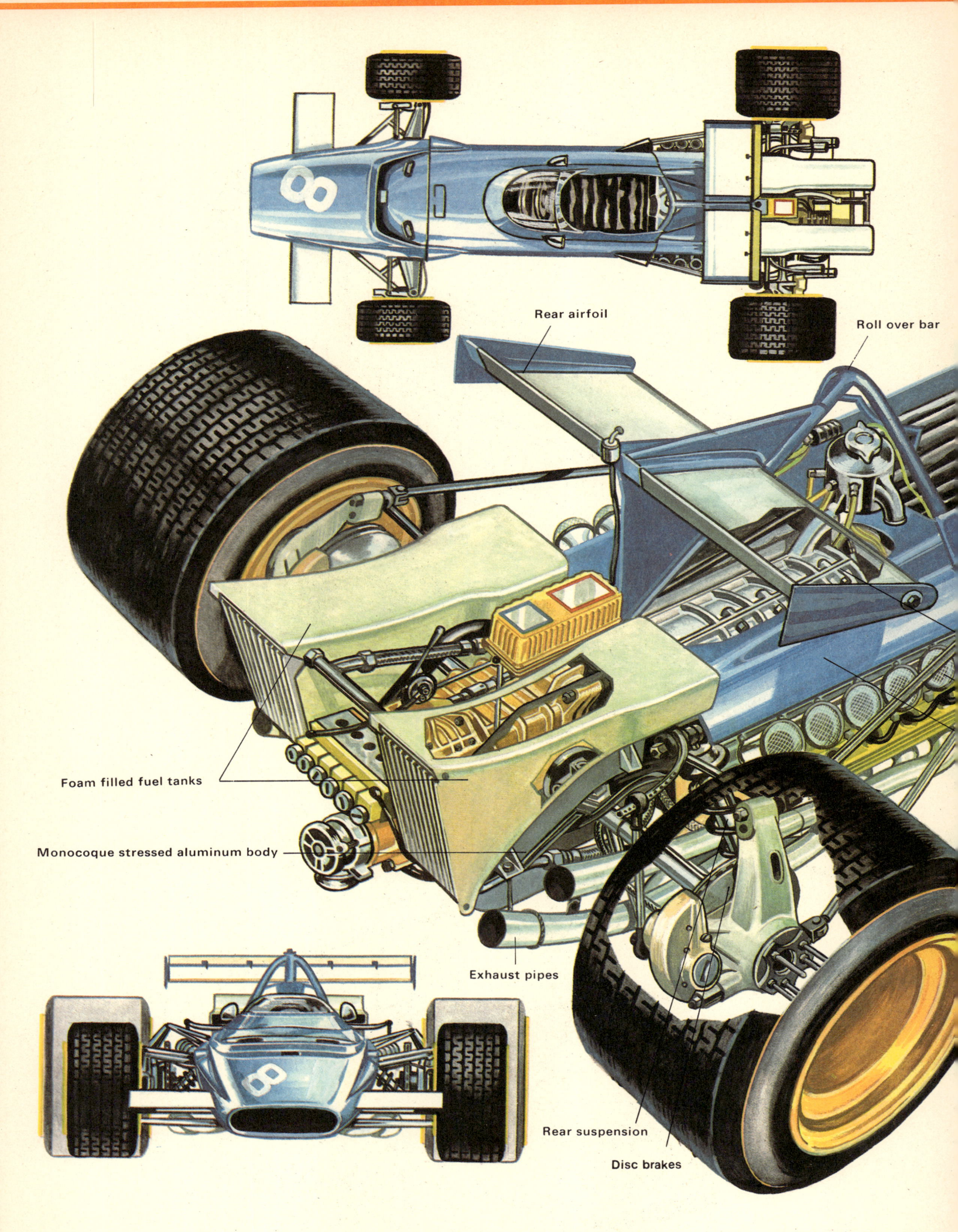

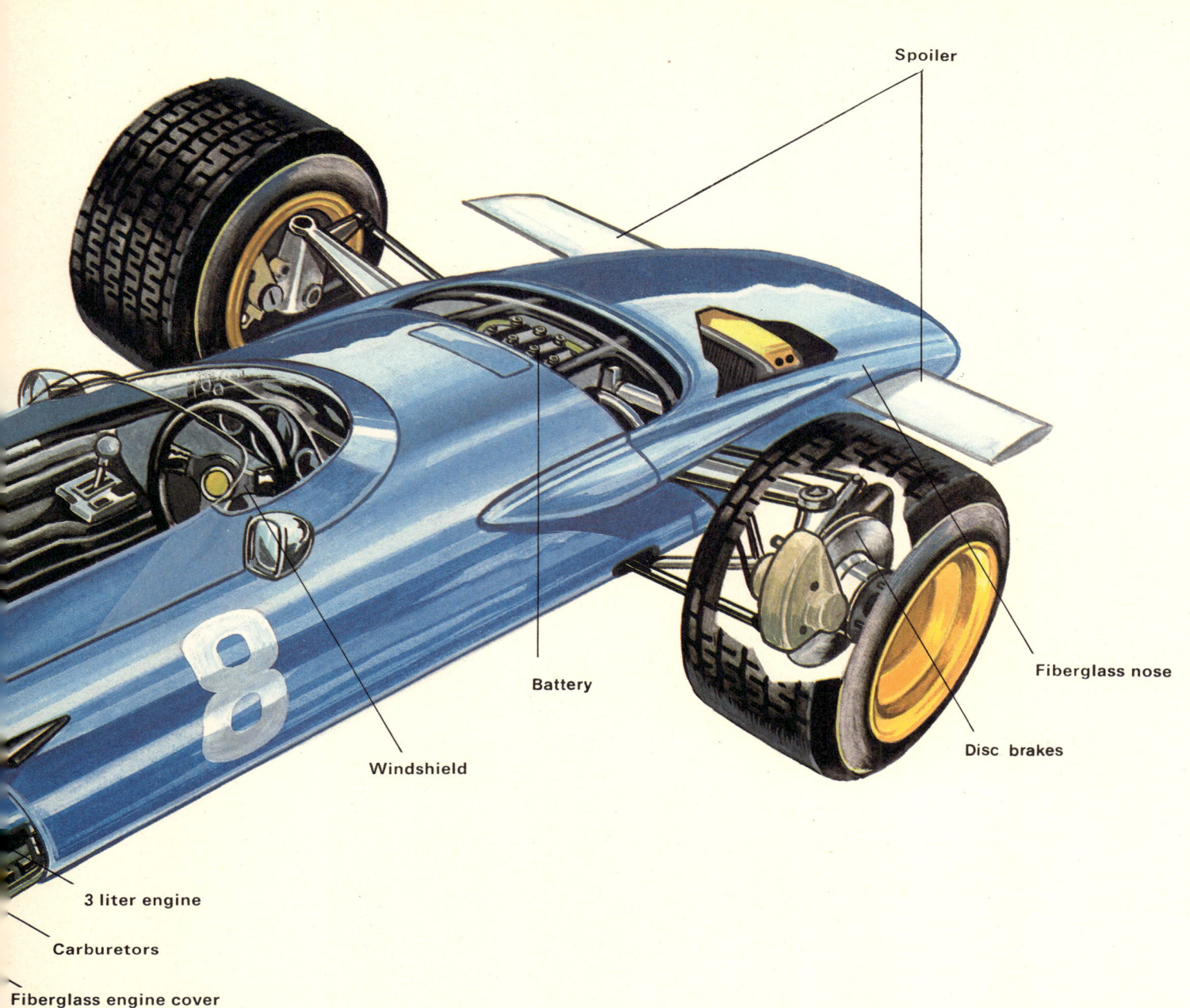

Fast racer: FORMULA ONE CAR

Motor racing is an international sport and the cars that compete must naturally conform to certain rules. This is known as the formula. There are several formulae, including Formula Four, Formula Three, Formula Two and Formula One.

In the last fifteen years Formula One cars have changed almost beyond recognition. The traditional front-mounted engine has given way to the rear-mounted type, and most cars now adopt the monocoque type of body construction. The lower center of gravity, wide profile tires and adjustable front and rear airfoils all contribute to better road-holding, and speeds have risen accordingly. With a modern V8 engine typically producing 450 bhp, top speeds of over 200 miles per hour (322 kmph) are possible.

Formula One cars are limited to an engine capacity of 3 liters and in recent years one engine has completely dominated the racing scene — the Ford Cosworth V8. This engine is used by many chassis manufacturers, including such names as Tyrrell, John Player, Lotus, Yardley Maclaren and Hesketh. Other engines, such as the Ferrari Fiat 12 and the Matra V 12 are used only in their own manufacturer's chassis.

In such a potentially dangerous sport, safety regulations must of course be very stringent. External battery isolation switches must be provided, so that marshals may isolate the electrics in case of fire. The cars must have built-in extinguishers, and the drivers wear full harness safety belts. Fuel is carried in a double skinned foam-filled tank which forms part of the monocoque body.

Weather conditions can have a significant effect on high performance racing cars, and this is especially evident in the behavior of tires. Generally two types are available for racing: dry weather tires, which are almost smooth with no tread pattern at all; and wet weather tires, which have a more familiar tread pattern. In the event of a sudden change in the weather, the thirty seconds or so spent changing the wheels in the pits can be made up several times over in the course of a Grand Prix race.

Space-age giant: CRAWLER TRANSPORTER

The great rockets launched from Cape Kennedy are assembled and prepared for their voyages into space three-and-a-half miles (5.63 km) from the launch pad. The problems of moving a Saturn V rocket of nearly 3,000 tons, its mobile launcher of 6,000 tons and the mobile service structure of nearly 5,000 tons to the launch pad were enormous. A search for a suitable vehicle centered on a Bucyrus-Erie mechanical shovel, weighing 9,000 tons. It was this vehicle that was adapted to a new space-age role. Two crawler transporters were built, at a cost of around $12 million for the pair.

The crawler has two operating centers — driving cabs — at diagonally opposite corners of the vehicle, controlling the movement, leveling and equalizing of the load. The controls allow the technicians to compensate for winds of up to 68 knots pressing on their vertical load. Driven by 16 motors in all, four to each crawler unit, the crawlers can pivot so that even a crabwise movement is possible in maneuvering the load. Disc brakes on the drive shafts enable the vehicle to stop smoothly, and if there is a breakdown in any of the motors the brakes apply automatically. Despite the gigantic scale of the crawlers, they inch their way to their destinations quietly. The two parts of each vehicle, chassis and platform, rest on steel link trackbelts like those of a tank, but each link weighs around one ton.

Crawler transporter

Chassis: 131ft long, 114ft wide, 20ft high
    (39.92m x 34.75m x 6.1m)
Speed: unloaded 2mph (3.22kmph); loaded 1mph
    (1.609kmph); climbing the 5° ramp to the
    launch pad ½mph (0.805kmph)
Weight: 3,000 tons
Load: 6,000 tons
Powered by 1,000kw DC heavy duty electric
    generators driven by 2,750hp diesel engines

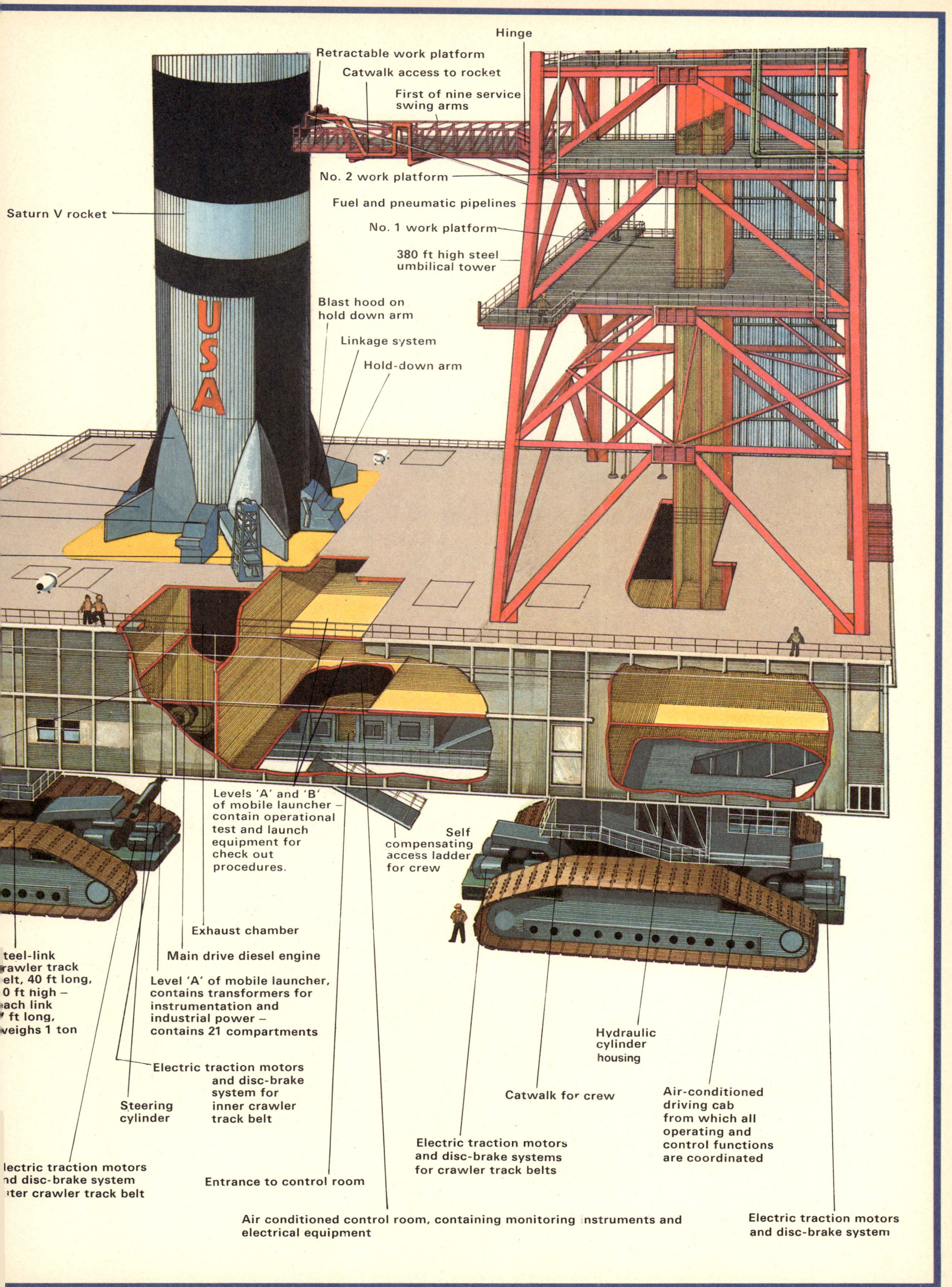

Gathering the grain: COMBINE

Since about 1935, the back-breaking and time-consuming jobs of reaping and threshing grain can be carried out by one man with a combine. Man and machine can harvest and thresh up to four acres an hour, collecting eight tons of grain in a high-yield region. An internal hopper holds up to four tons, and the grain can be unloaded into a truck driven alongside without stopping the harvester.

Rotating sails up to 20 feet wide pull the crop downwards. The cutting bar consists of knives which move quickly backwards and forwards, like an electric hedge clipper or a hairdresser's clippers. They chop off the crop at stubble height, and sensing fingers beneath automatically raise and lower the cutters to follow the shape of bumps in the ground. A coarse-pitched rotating screw, called an auger, sweeps the cut crop to the center of the table, where a conveyor lifts it up to the threshing drum. This is a high-speed cylinder whose beater bars knock the grain out of the heads of the crop. Screens remove large pieces of chaff, and an air blast winnows the corn by blowing away small pieces of husks and straw. The main bulk of the straw passes back from the threshing drum along toothed *straw walkers* (also called a jog trough) and may be chopped before falling to the ground behind the moving machine. The separated grain is moved across by another auger and up to the top of the storage hopper by a chain conveyor.

The whole machine is driven by a diesel engine delivering up to 150 horsepower. The various rotating parts are driven by belts, and the cutting mechanism is raised and lowered hydraulically. The driver, who steers the combine by the back wheels, may have an air-conditioned cab to protect him from dust and heat.

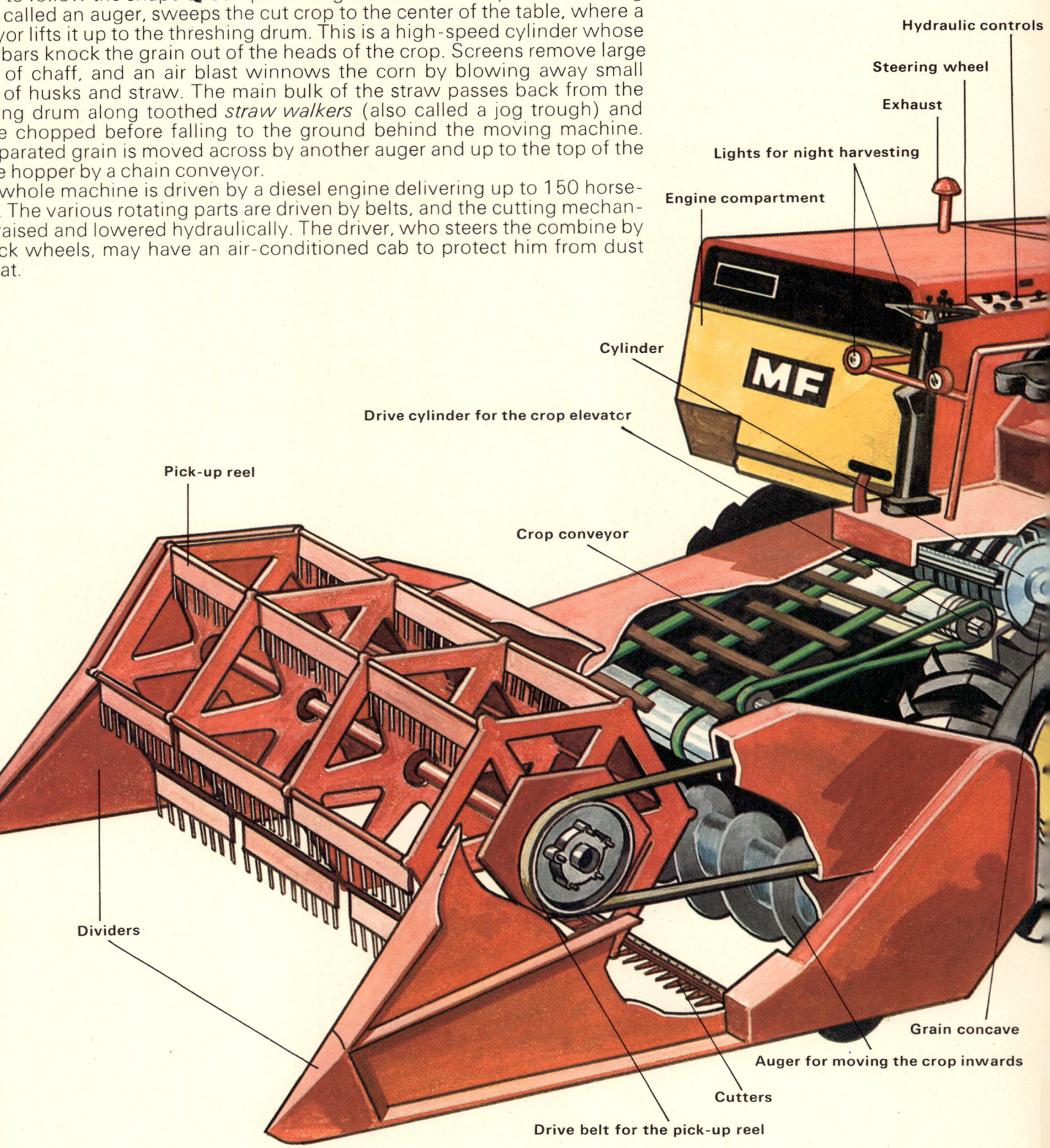

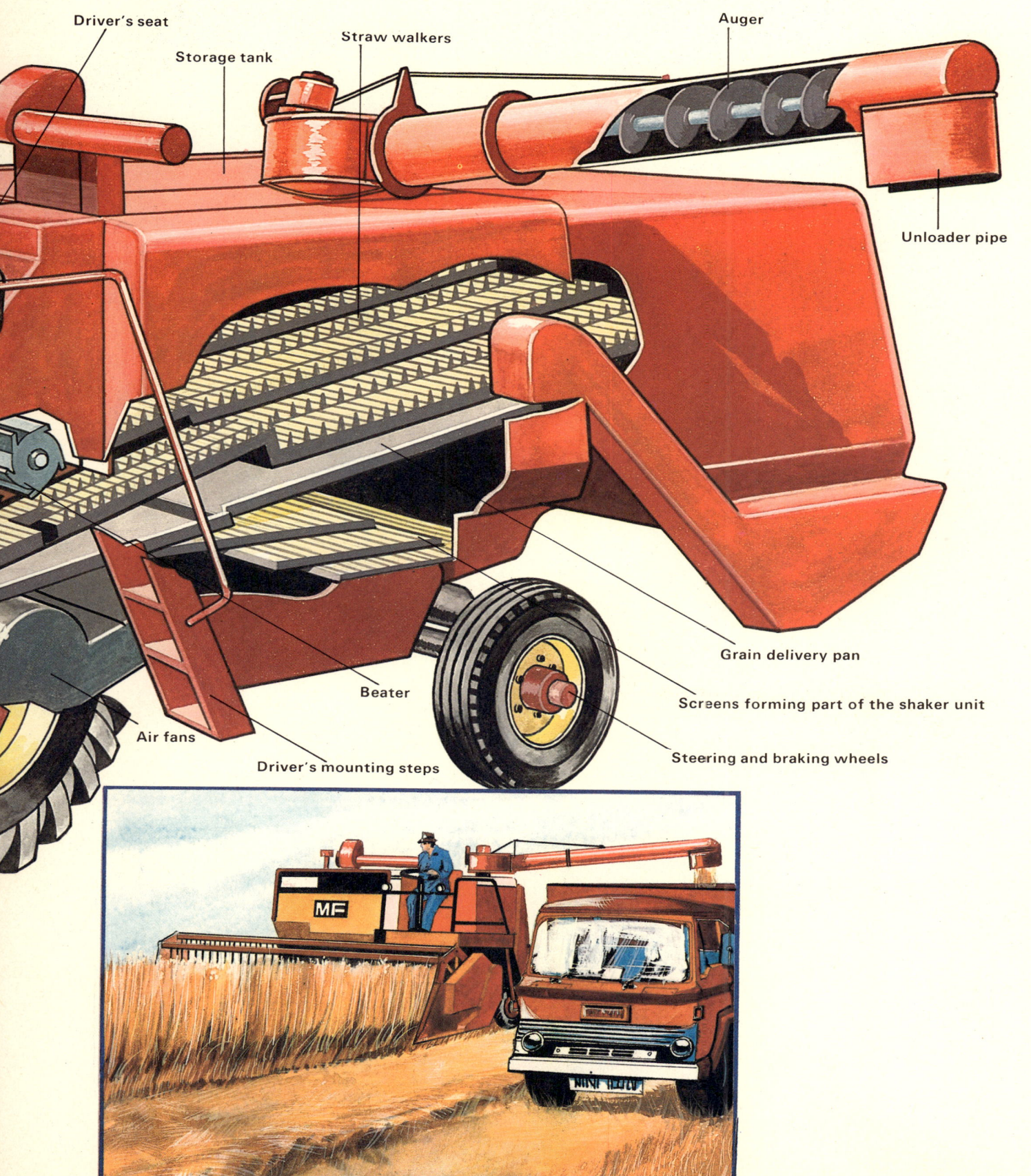

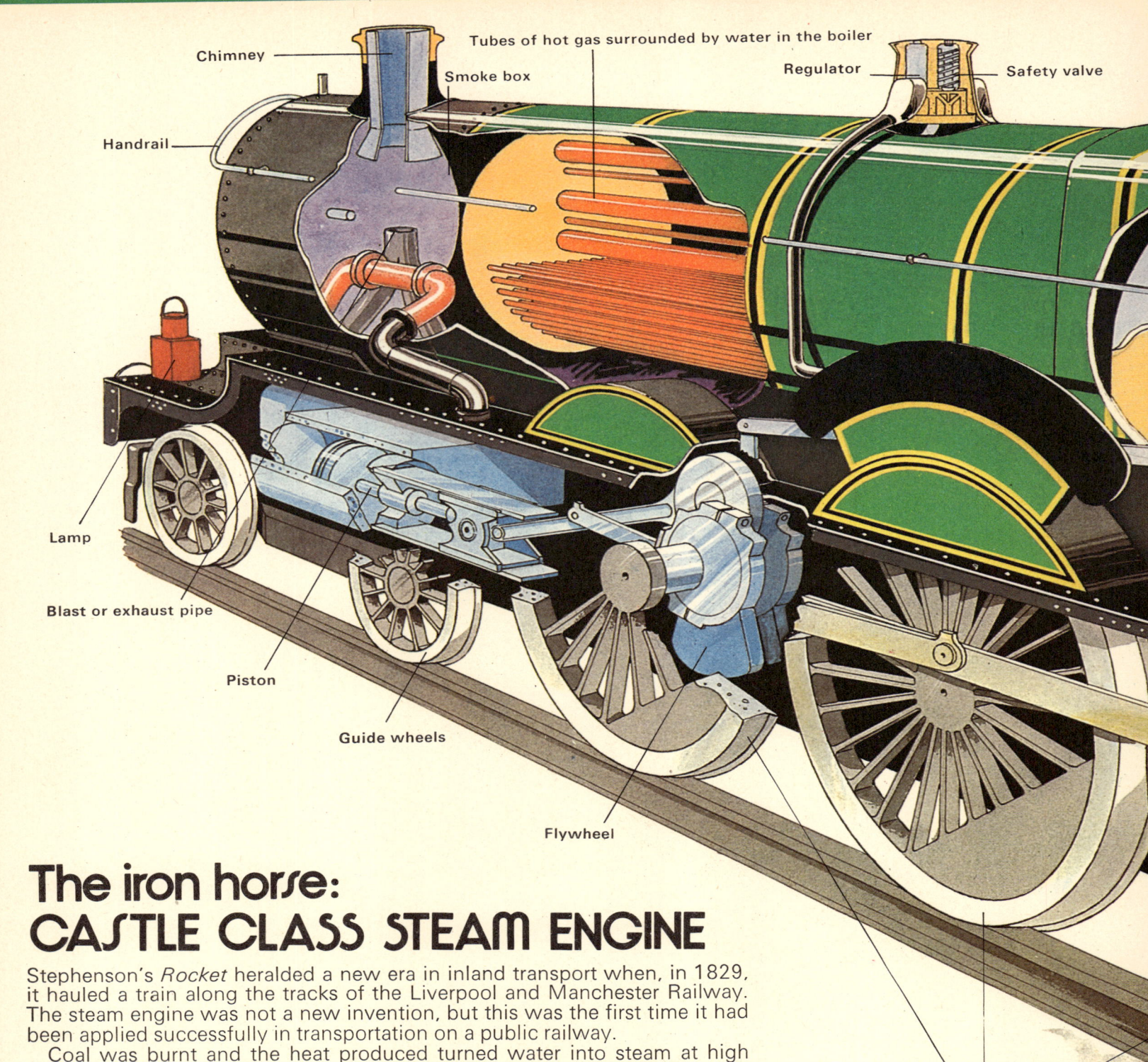

The iron horse:
CASTLE CLASS STEAM ENGINE

Stephenson's *Rocket* heralded a new era in inland transport when, in 1829, it hauled a train along the tracks of the Liverpool and Manchester Railway. The steam engine was not a new invention, but this was the first time it had been applied successfully in transportation on a public railway.

Coal was burnt and the heat produced turned water into steam at high pressure. This was then admitted to a cylinder, by way of a valve arrangement, and being at a high pressure forced the piston down the cylinder. More steam was then fed into the other end of the cylinder via second valve mechanism and thus forced the piston back again. The process was repeated and the lateral motion of the piston could be converted into circular motion of wheels by various arrangements of cranks.

From 1829, the competition between private railway companies became very intense, as they fought for the patronage of would-be travelers. This resulted in many fine steam engine designs being produced. Typical of the masterpieces of the era was the *Castle* Class of locomotive designed by C. B. Collett, who was chief mechanical engineer to the Great Western Railway in England from 1921 to 1941.

The Castles were introduced in 1923 and were among the finest steam locomotives ever to be built. The engines had four cylinders, a water capacity of 4,000 gallons (15,140 liters), a coal capacity of 6 tons and developed a boiler pressure of 225 lb per sq in. (15.8 kg per sq cm). The working weight of the engine was 126 tons and the driving wheel diameter was 6 feet 8½ inches (2.03 meters). The Castle Class locomotives had a 4-6-0 wheel arrangement and were intended for use on high speed passenger services. Top speed was approaching 100 mph (150 kmph).

Some of the British trains pulled regularly by engines of this class included the Cornish Riviera Express and the Cheltenham Flyer, and it was on this train that the fastest ever run from Swindon to Paddington was achieved in 1932. The engine Tregenna Castle traveled the 77 miles at an average speed of over 80 mph (120 kmph).

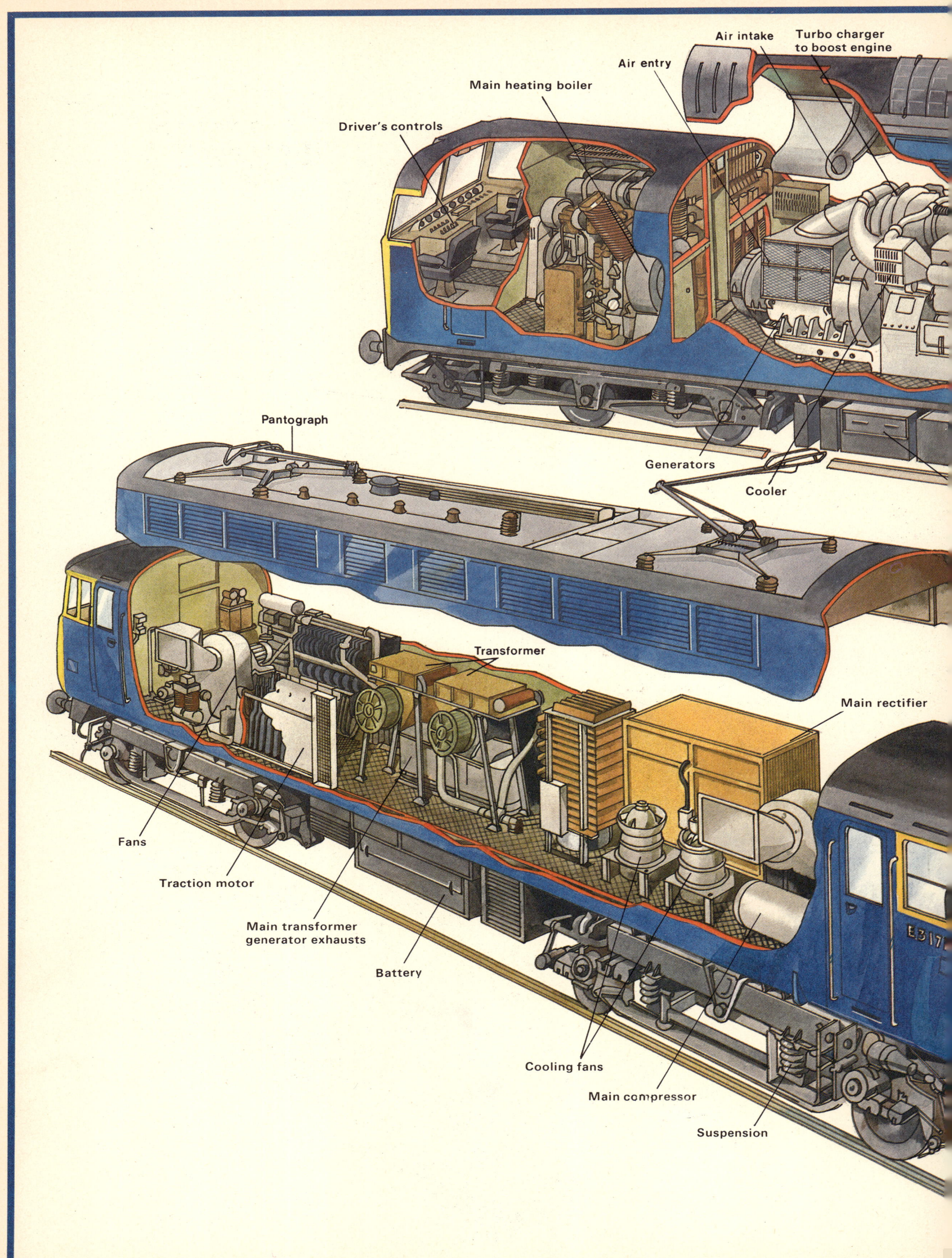

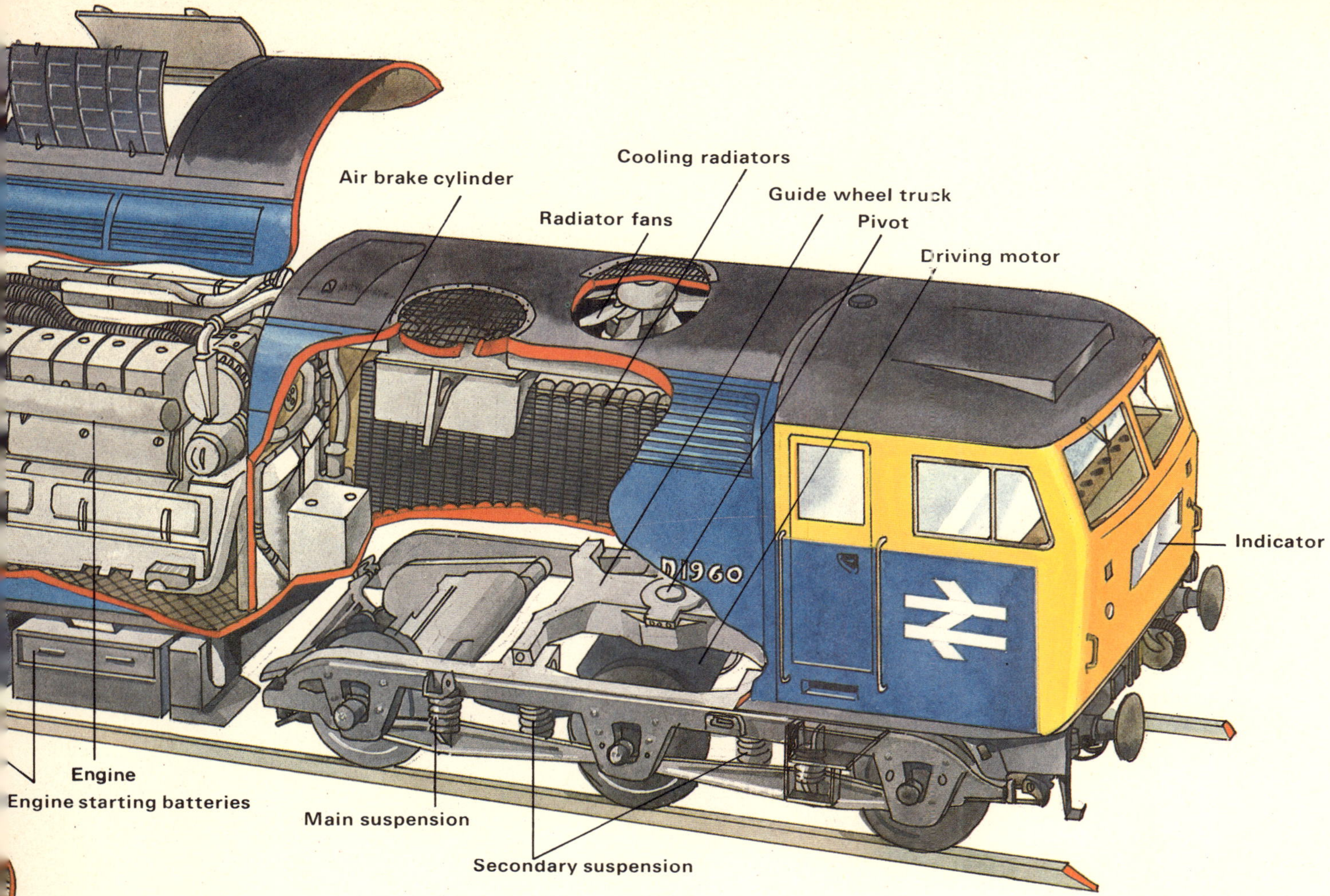

On the rails: DIESEL AND ELECTRIC LOCOMOTIVES

The electric locomotive has its origins in the late nineteenth century. It was used in situations where steam locomotives were unsuitable — in subway railways, mountainous areas and mines. Since the 1950s most electrification of railway systems has been based on the 25 kilovolt, single phase AC 50 Hz overhead-collection system.

DC motors are simpler to control than their AC equivalents, and are preferred for transportation systems, so it is necessary to include rectifying equipment in the locomotive. The alternating current is collected by the pantograph, transformed down to a lower voltage, and then rectified to produce direct current. In the AL5 type of locomotive, shown here, each of the two four-wheeled trucks is driven by its own motor. The maximum speed produced by these engines is about 90 mph (145 kmph).

Diesel electric locomotives carry their own energy source with them. This is in the form of diesel oil, which is used to power a diesel engine connected directly to an electric generator. This provides the power to drive the DC electric motors, which are mounted on the trucks. The advantage of this system is that it avoids the problem of collecting the electricity from outside the locomotive, either by way of a third rail or a pantograph.

A disadvantage lies in the necessity of carrying a large quantity of inflammable fuel, and it is essential to fit automatic fire-fighting equipment that is triggered by temperature sensitive switches. The added weight of the diesel engine and electric generator makes the diesel locomotive heavier than its purely electric counterpart, which leads to a lower maximum speed.

Inter-city: HIGH SPEED TRAIN

Over the past ten years, railway engineers have recognized that the limit on safe and economic speed of trains running over today's track is much higher than the 100 mph (160 kmph) previously loosely considered as the highest possible; though on occasions special trains run over cleared track with no thought for cost, had already reached 205 mph, (330 kmph). A pioneer in this work was British Rail's Research Center at Derby, where fundamental research gradually solved the problems of how railway wheels and vehicles behave.

While the Japanese built new trains capable of running at 130 mph (210 kmph) over completely new track, without junctions or sharp curves and powered by no less than 12,000 horsepower in each train, British Rail learned how to make trains run much faster over today's track, with all its curves, junctions and other hazards. The outcome is a new race of 145 mph (235 kmph) freight vehicles and an Advanced Passenger Train (APT) able to cruise at about 155 mph (250 kmph), powered either by groups of gas tubines or by overhead electric supply. One feature of the APT is its extreme lightness, so that acceleration will be higher and air drag less, thus cutting the cost of the energy consumed. Another is that passenger comfort is improved by gently tilting the coaches as the train streaks round bends. As a first step British Rail has put into service a slightly less radical HST (High-Speed Train), which does not have tilting bodies but nevertheless represents a major advance in power/weight ratio with improved comfort. At each end of the HST is a lightweight streamlined power car housing a 2,250 hp diesel engine and a modern cab. Between them can be coupled any desired number of passenger cars, seven being a common number. The first HST reached 150 mph (241 kmph) on the Eastern Region in 1974, and production trains on the Western Region cut journey time between London and Bristol from 98 to 70 minutes, despite being held to 100 mph (160 kmph) except over one short section of track. Unlike conventional trains, they can run at top speed for nearly the whole of a long journey.

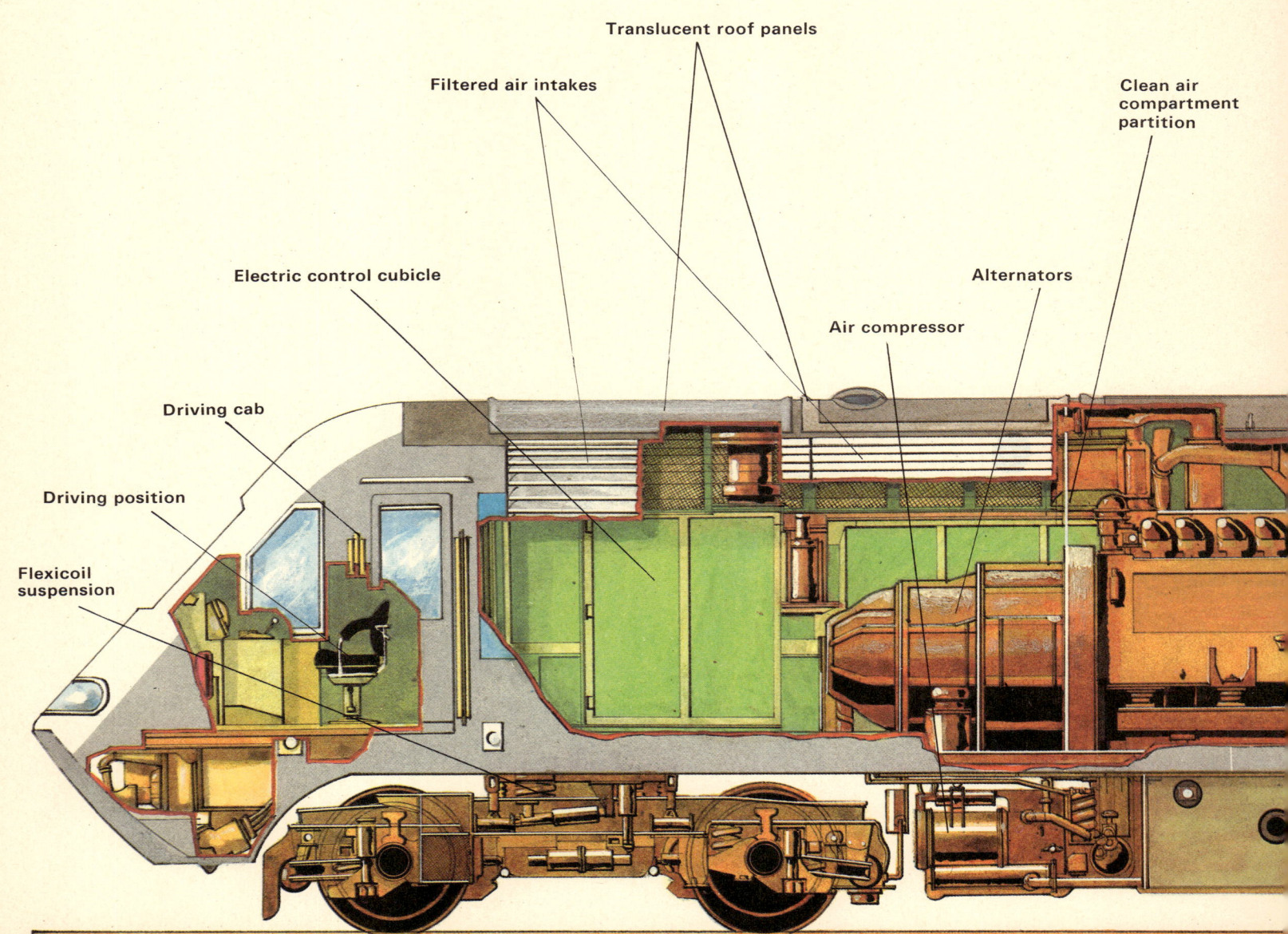

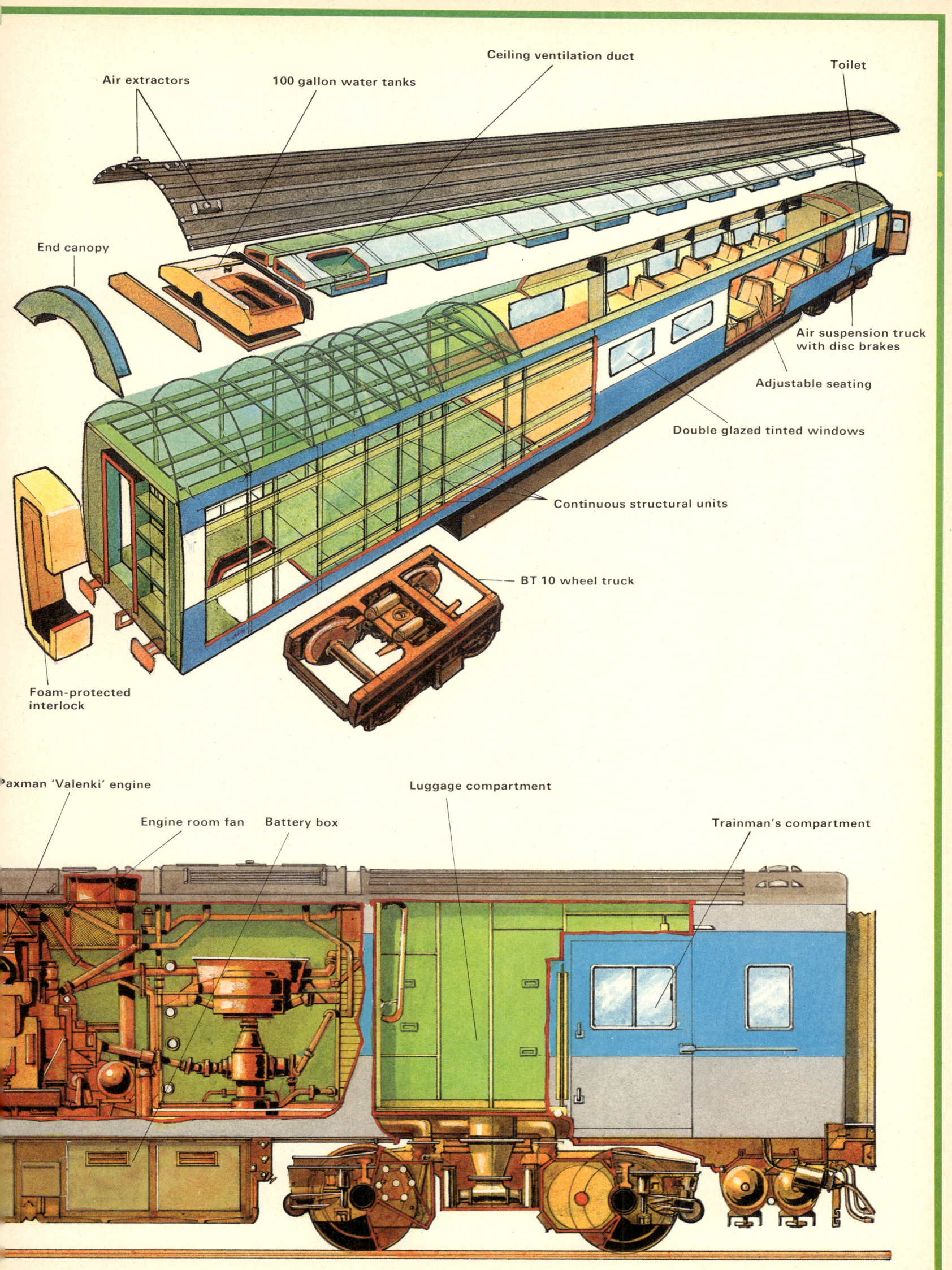

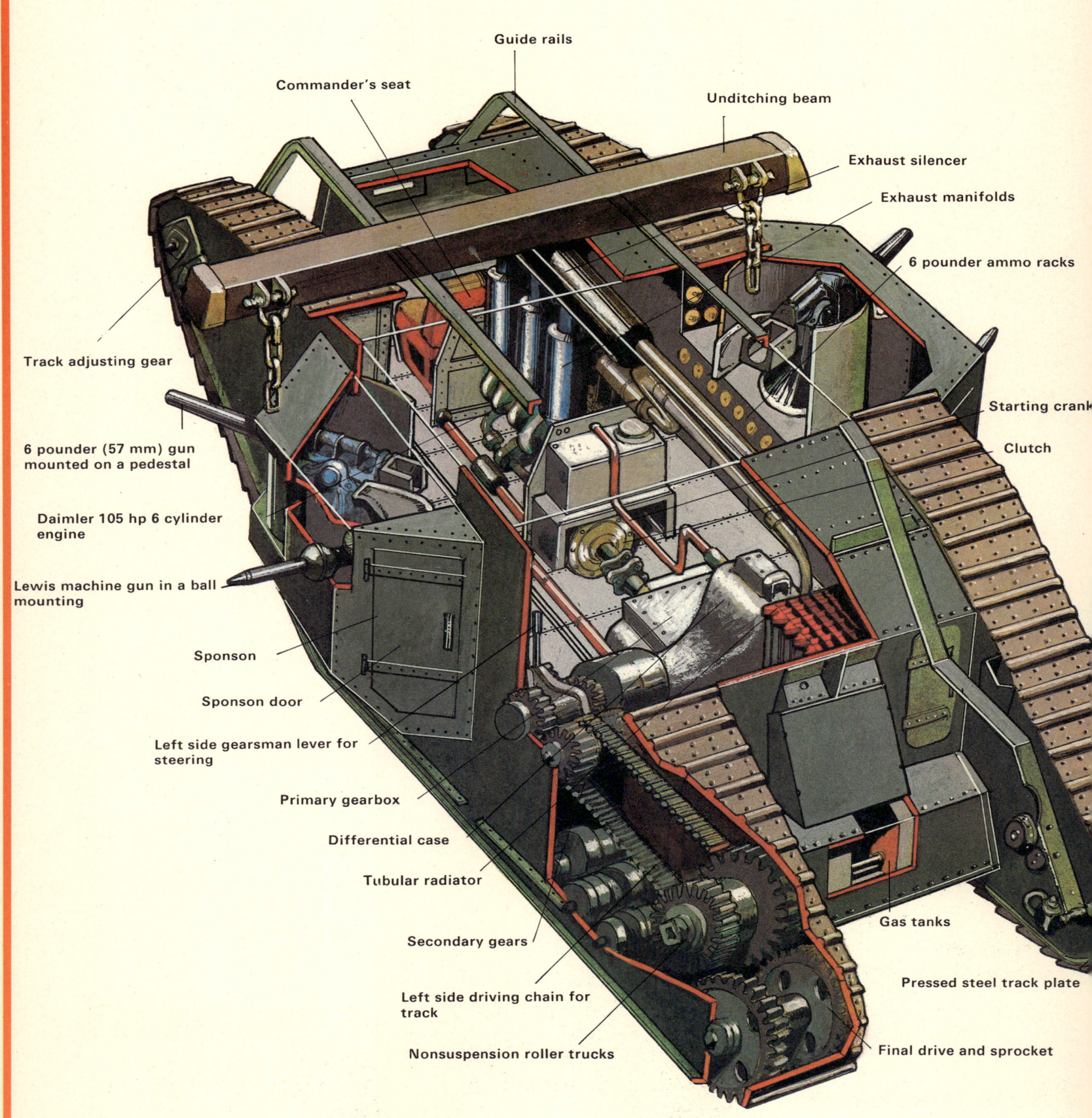

Armored to battle: MARK IV AND SHERMAN TANKS

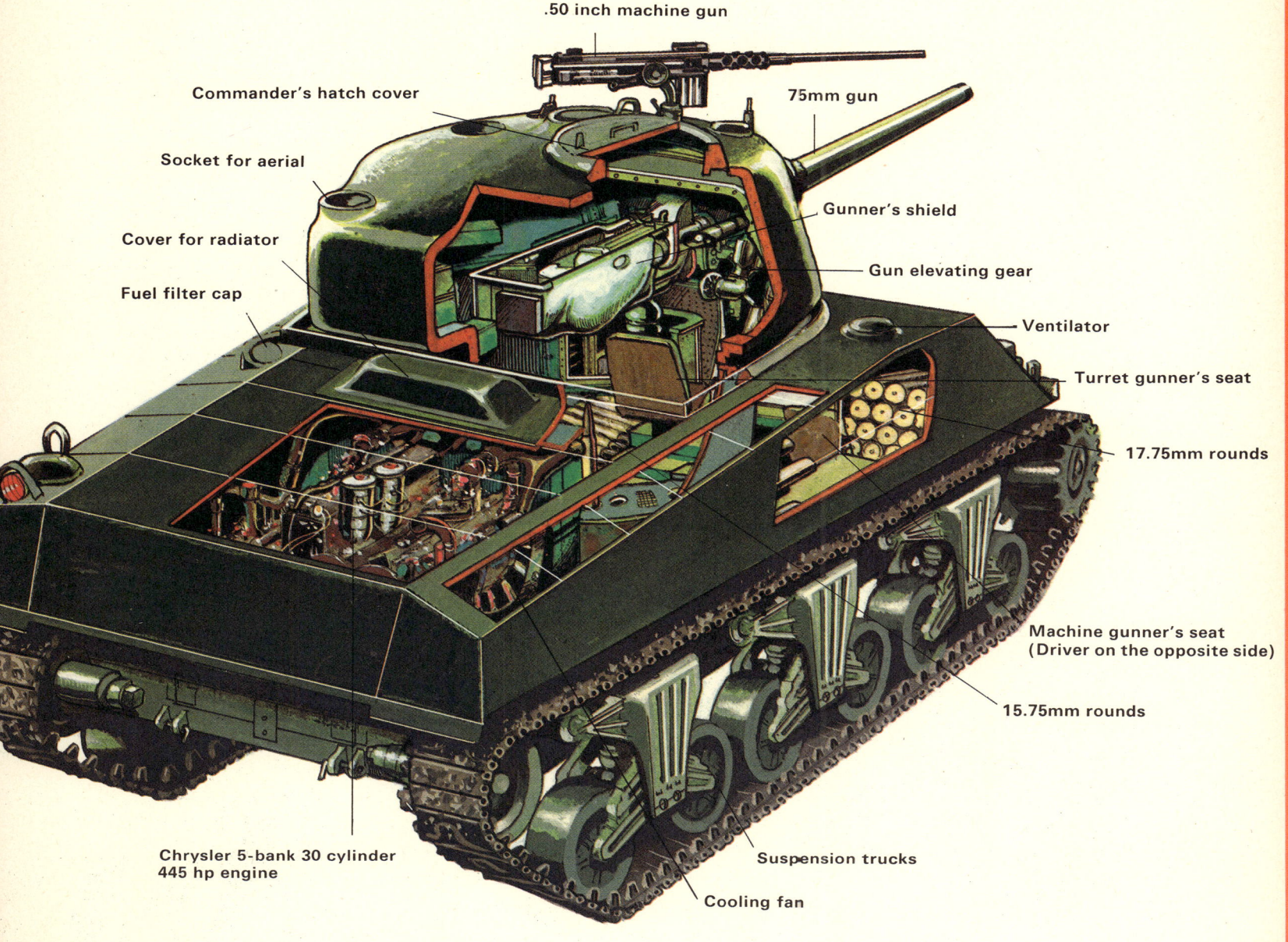

The British tank Mk IV

Tanks, or armored fighting vehicles (AFVs), made their first appearance on the battlefield during World War I. Their development during the early years of the war was naturally a very hush-hush affair, and their code name was "water tanks for Russia". Afterwards part of the name stuck and is still used today.

The first really successful tank was the British Mark IV, despite its weight and poor maneuverability. It used a Daimler 105 bhp gas engine but its 28 ton weight limited its maximum speed to less than 4 mph (6 kmph). The tank was manned by a crew of eight – a commander (who also operated the brakes and a machine gun), a driver, four gunners and two gearsmen.

Two types were made: the *male* version, which had one six pounder and one machine gun in each sponson, and the *female* version, which had two machine guns in each sponson. Both types had an additional front mounted gun which was operated by the commander. The armor plating had a maximum thickness of $\frac{15}{32}$ inch (12 mm).

The M4 General Sherman

The Sherman tank was one of the most dominant of the Allied Forces' tanks in World War II. Introduced in 1942, it had a 75 mm gun mounted in a fully rotating turret. A new feature of the Sherman was its gyro stabilized gun mount, which allowed it to fire accurately even when on the move. It was able also to fire either armor piercing or high explosive ammunition.

In contrast to the lumbering tanks of World War I, the Sherman's 500 bhp Ford engine gave it a top speed of 26 mph (40 kmph). Its maximum armor thickness of $2\frac{61}{64}$ inches (75 mm) contributed towards its 34 ton weight. A crew of only four men was needed to man it efficiently.

The Sherman was so successful that many variations in design were introduced. The basic chassis was used for recovery vehicles, gun carriages, anti-aircraft gun vehicles and rocket firers. In all, some 42,000 Shermans or variations of them, were produced.

Mechanical digger: POWER SHOVEL

In the great age of railroad building, towards the end of the last century, most of the work was done by gangs of stalwart workmen using picks and shovels, occasionally assisted by explosives. The surge in civil engineering projects over the last few decades placed heavy demands on manpower so engineers worked to replace the great gangs of men with their hand tools. They came up with a variety of machines for digging and clearing ground.

Most of the pick and shovel work is now handled by a few men operating such machines as bulldozers and mechanical shovels. These teams with their mechanical aids can move great masses of material and work at speeds that would have shaken the unskilled laborers of seventy years ago.

The mechanical excavator on this page is one of the machines that makes possible the rapid building techniques of today. On its chassis the manufacturer can mount a mechanical shovel, dragline excavator, a grabbing crane or a lifting crane, making it a thoroughly versatile machine. While, at 78 tons, it is not the largest of its kind, it can take an impressive three-and-a-quarter cubic yard ($4\frac{1}{4}$ cubic meters) mouthful of rubble at a bite. It will pivot on the swing circle (shown in the illustration below) so that the machine will operate in awkward and confined spaces. The caterpillar tracks provide a firm base for the working structure, and extra large caterpillar tracks can be fitted to cope with severe conditions. The machine is powered by a diesel engine, in this case an eight-cylinder engine producing over 200 shaft (brake) horsepower.

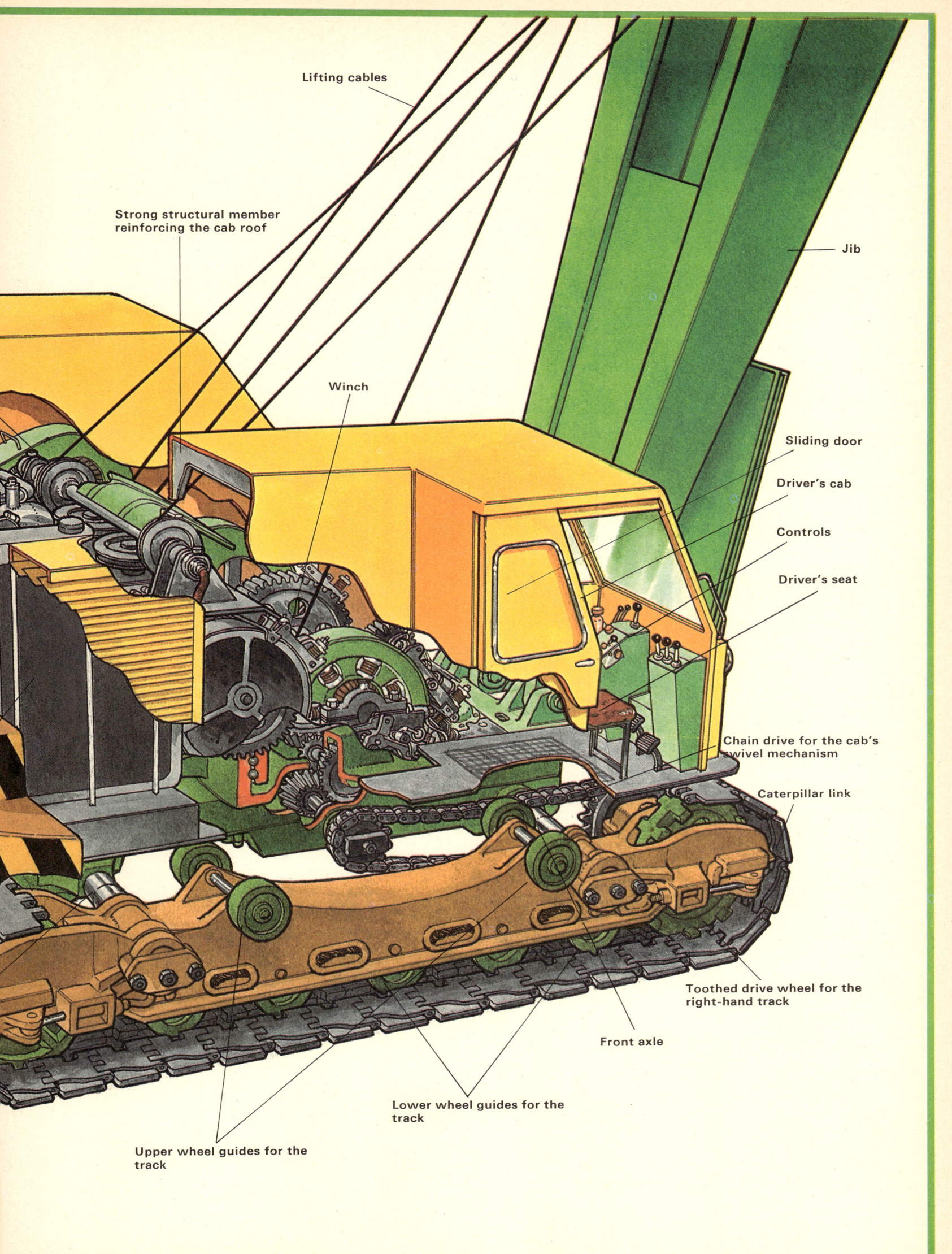

Savannah

Britannia

Tolchester

Elbe

Admiral Graf Spee

Enterprise nuclear powered aircraft carrier

ON THE WATER

Man, although he is a land animal, inhabits a world whose surface is mostly covered with great oceans. Early in his history, primitive man used rafts and hollowed logs to pole and paddle about the lakes, marshes and rivers of his hunting grounds; but his sense of curiosity grew, and he began to attempt ocean voyages. Fear of the vastness of oceanic distances held him to coast-hugging exploration for many centuries, while generation after generation of experience added to the sea lore that could be handed down.

With experience, mariners developed a sixth sense for direction. They learned to adjust their passage by observing fixed stars, the movement and angle of the Sun and winds, and by *reading* the currents. It was experience won the hard way, by shipwreck, starvation and thirst. Gradually the boundaries of the known world were pushed back and the technology of ship-building grew more important to the intrepid voyagers on their longer and longer journeys.

By the second millenium BC Mediterranean peoples, such as the Egyptians, were building sea-going vessels with sails that we would recognize as seaworthy ships for coastal voyages. The earlier generation of rafts, in spite of their simplicity enabled the remarkable Polynesian sailors to cross the South Pacific and populate New Zealand and later Hawaii. The craft of the ship-builder grew more accomplished, and bold experiments led the shipwright to build flexible hulls of planks secured with hooked nails to the timbers of the frame. The planks, bound together with thongs or ropes and tarred, flexed under the battering of the waves but did not break or let through much water. It was hulls like this that carried the daring Norse sea rovers around the Baltic, across the North Sea, down to the Mediterranean and across the Atlantic. Their strong arms powered the ships, but the oar would give way to the sail when the wind was favorable.

The sail became the means of harnessing the wind's power for the sailor, and continued as the source of power at sea through the great period of exploration during the fifteenth and sixteenth centuries, when Man learned the true plan of the world he inhabited. The square rig of Western ships was developed by adding more masts to allow space for more yards on which to hang yet more canvas. Sailors observed the flexible use of the lateen sail, favored by Mediterranean sailors, and added it to their own square riggers.

From the beginning of the nineteenth century, the challenge of powered ships focused the attention of builders of sailing ships on speed. Over the next 80 or 90 years, they produced the fastest and most graceful ships ever seen just as they were forced to give way before the growing effectiveness of powered ships.

The Age of Discovery in the fifteenth and sixteenth centuries advanced the skill of navigation as a developing Europe sought cheap raw materials and fresh markets for her goods. Almanacs and instruments were devised to ever increasing standards of accuracy, and the work of the instrument maker became an art as well as a craft. It was a time when it was thought appropriate to give a fine instrument a shape and decoration that was beautiful as well as practical. If a sextant worked well, the craftsman would celebrate the fact in fine etched decoration and embellishment.

Surveying and mapping of the coasts, shoals and currents proceeded fast and, like the instrument makers, cartographers adorned their maps with fanciful designs — mermaids, fabulous monsters and treasure hordes.

The industrial revolution in Western Europe brought together the energies of engineers and factory owners to provide large, fast ships to carry their wares to new colonies and expand their share of the world's markets. The new iron hulled ships of Isambard Kingdom Brunel, the steam powered paddles and the still newer screw driven ships opened a new era for sailors. The days of the great iron ships with powerful engines fed by sweating stokers with coal had all arrived by the mid-nineteenth century, and graceful sailing ships were making their last brave fight for commercial survival. The weapon they chose was the fast and beautiful clipper ship.

At first, the steam ship required so much coal that accommodation of passengers and cargo was severely restricted by the space needed for storing fuel. As engineers built more efficient engines the problem was resolved. The discovery of yet another fuel and ways of using it soon led to the development of diesel engines for ships. The space required for storing the fuel oil was relatively small, and the diesel powered cargo ship slowly edged ahead of the steam powered cargo ship.

The loading and unloading of ships was, and still is, the most wasteful part of their working lives. The packing of cargoes in large containers was attractive to manufacturers and land transport organizations, but the large containers were awkward to load through hatches into the holds of ships. The solution was to design ships specially for container cargo. The container ship and the specialized dock handling arrangements are now installed in many ports around the world, shortening the wasteful period in port.

New types of ships have appeared on our coasts in the last twenty years. The hovercraft is part land and part sea craft, and even in a sense part aircraft. The ferry service that it provides is especially valuable to points where there are no normal docking facilities. It simply rides over the waves, up the beach and deposits its passengers on the shore. The fast hydrofoil makes swift, smooth river and lake trips, but has problems with the rougher surface of the sea. The newest power source, atomic fission, is now installed in several ships, but cannot yet be widely used because of difficulties in persuading port authorities of the safety margins in the ships' designs.

Men have forced their way across the sea for thousands of years, but it is only in this century that they have successfully crossed the seas under the surface. The nuclear submarine can even circumnavigate the globe without surfacing. As yet, the only non-military use of submarines has been for scientific research and as a tool for engineers. Man underwater has penetrated from about 100 feet (30.5 meters), that a sponge diver can manage without artificial aids, to nearly 36,000 feet (1,097 meters) in a bathyscaphe constructed to withstand colossal pressures to its hull.

Despite advancing knowledge of his environment, Man has very much to learn of the oceans that cover four-fifths of the surface of his planet. His knowledge of the seas lags behind his understanding of land.

Cars across the sea: CAR FERRY

The end of World War II opened Europe once again to the private motorist, and he demanded a means of crossing short sea routes with his car. At first, old tank-landing craft were used at several ports, but soon purpose-built ships gave a more dependable and comfortable service. As road transport recovered its importance and then challenged the supremacy of railways for carrying large loads over long distances, larger ferries were built. Companies demanded ships that simplified and speeded up the loading and unloading procedures. In the early ferries, vehicles were slung aboard by crane. This was a clumsy and slow method so designers returned to the tank landing craft principle of a ship which could lower her stern or her bows section or even both. This enabled drivers to take their vehicles aboard under their own power, reducing the turn-round time in port greatly.

The ferry below operates between the European mainland and Great Britain. It is designed primarily for the ferrying of large trucks, but also offers comfortable accommodation for some private motorists too.

Cerdic Ferry

Launch date: 1961
Gross tonnage: 2,455 tons
Length overall: 362ft 9in. (110.2 meters)
Beam: 54ft 3in. (16.6 meters)
Draft: 13ft (3.9 meters)
Service speed: 14 knots
Cabin berths: 38 including 1 two-berth deluxe, 4 two-berth, and 7 four-berth cabins
Vehicle capacity on main deck: 25ft x 40ft
Vehicle capacity on upper deck: 12ft x 40ft
Freight capacity in length: on main vehicle deck, 1,001ft 6in. (305 meters); on upper vehicle deck, 482ft 6in. (146 meters)

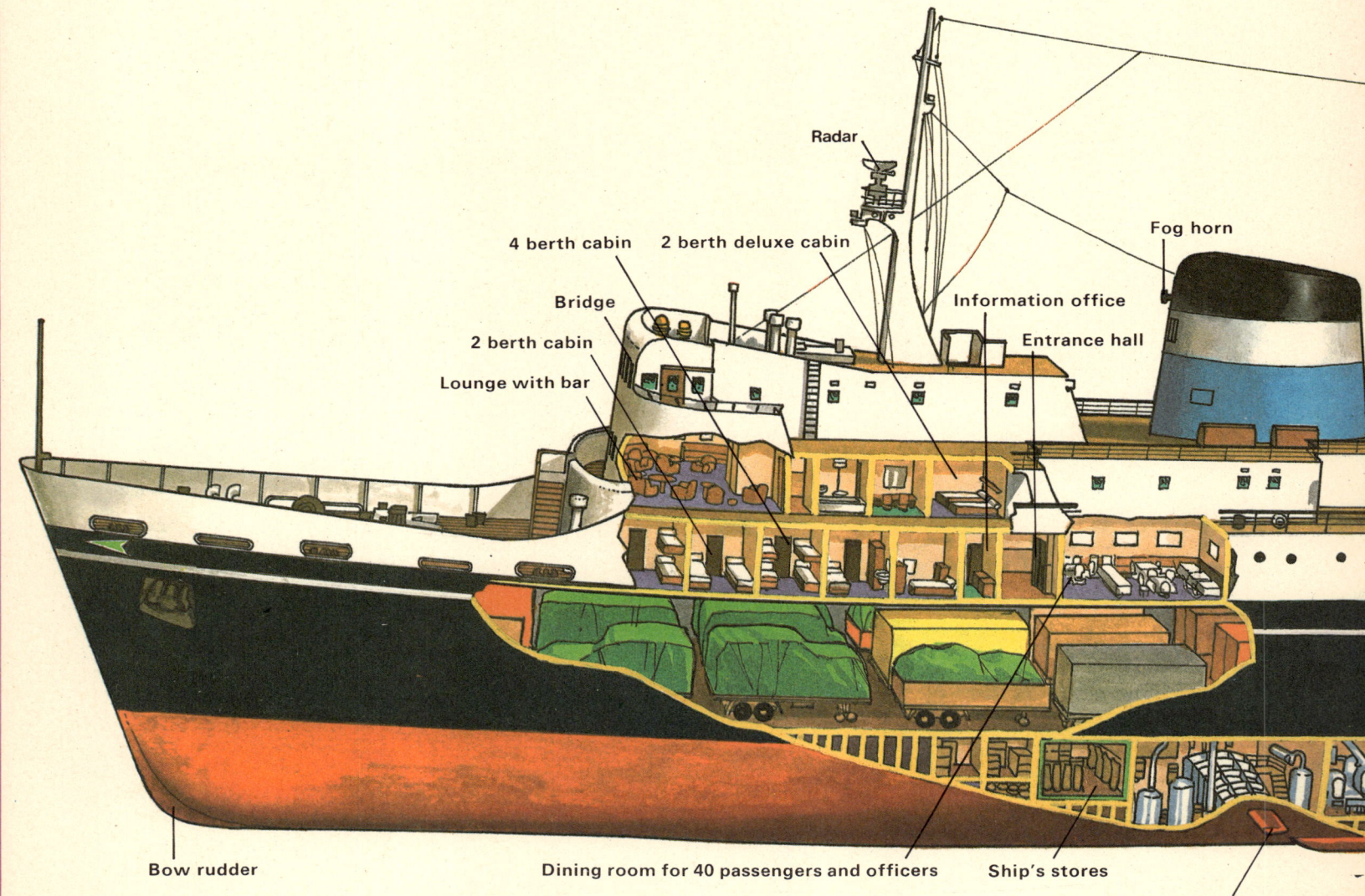

The roll-on, roll-off arrangement of this ferry is designed for a two-story system. A service road leads to a ramp across which light vehicles run onto the ship's upper deck. Heavy vehicles run through the stern opening from the dock level. In this way the weight is placed low in the ship, where it makes for stability.

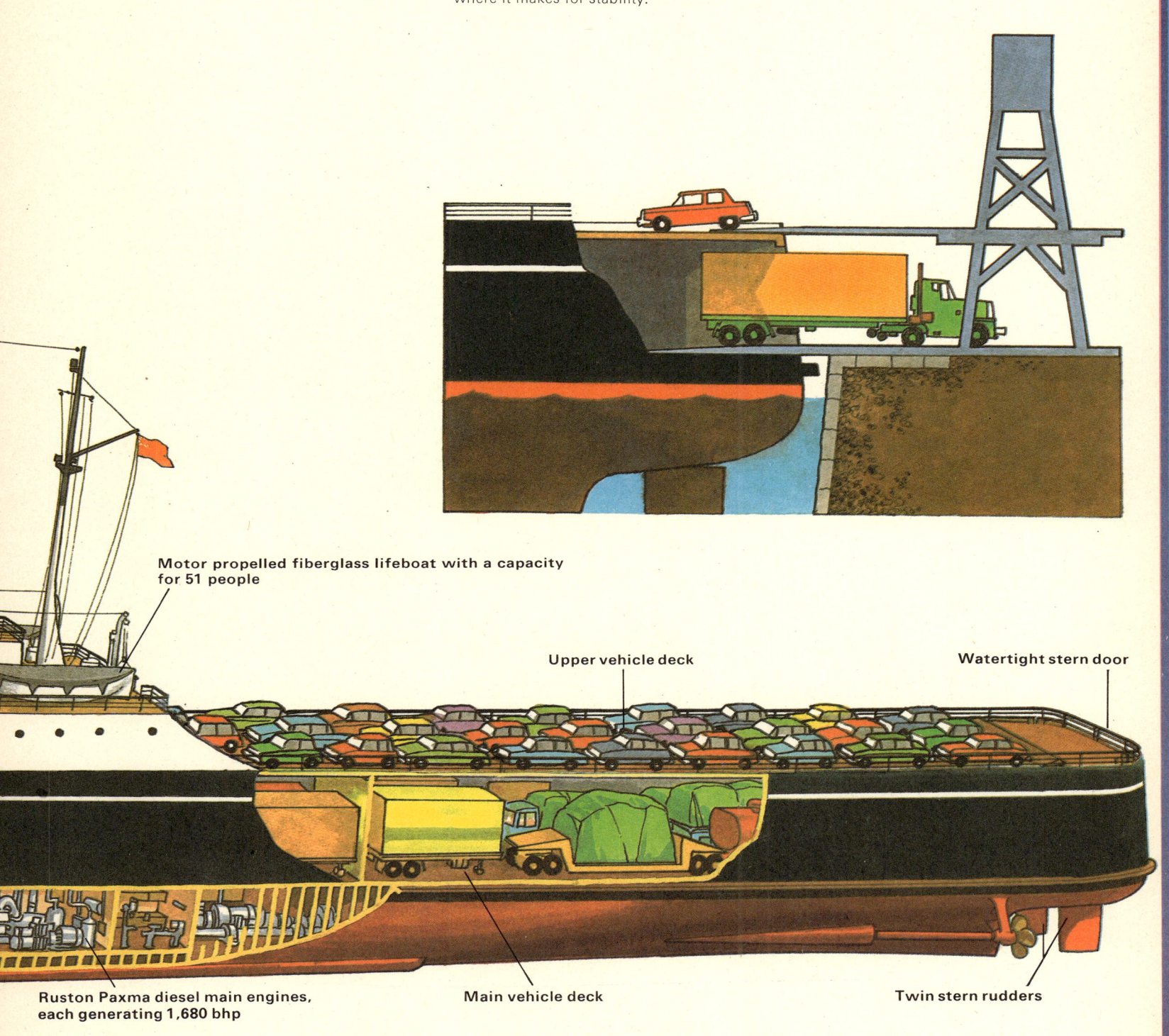

Boxed cargo: CONTAINER SHIP

The whole world of cargo shipping has been revolutionized by the introduction of the container ship. Previously, the holds of cargo ships were just huge spaces in which dockers would stack cargo — it might be as diverse as boxes of books, carcasses of beef, drums of paint, bags of cement and coils of wire. They found it difficult to keep the items separate, and in a rough sea they might become jumbled together and damaged. It was easy for some parts of the cargo — say a case of whisky or a box of cigarettes — to *disappear* at some point en route. Frequently loading and unloading took several days, because dockers lifted every piece of cargo through a hatch in the deck.

With the container, everything is packed at the start of the journey — long before reaching the dockside — in a large rigid box. This box has standard dimensions, measuring 8 feet wide, 8 feet high and 10, 20, 30 or 40 feet long (2.44 x 2.44 x 3.05 and 6.10, 9.14 or 12.19 meters). It clips securely onto the floor of a truck, railroad flatcar, freight aircraft or the deck of a ship. If necessary, several containers can be stacked one above the other, clipped together. All are locked before being sent so nothing can be pilfered. Longshoremen operating cranes with standard grabs that latch onto matching sockets in the containers' sides handle them swiftly. A load of thousands of tons can be stowed on board in half a day, filling the hold and the deck of the ship.

With the vast growth in container traffic has come a new class of specially designed container ships. Many are large, and nearly all resemble tankers in having superstructure and machinery at the rear of the ship. Most are fast, with speeds of a little over 25 knots. Propulsion in this ship is by STAL-Laval steam turbine

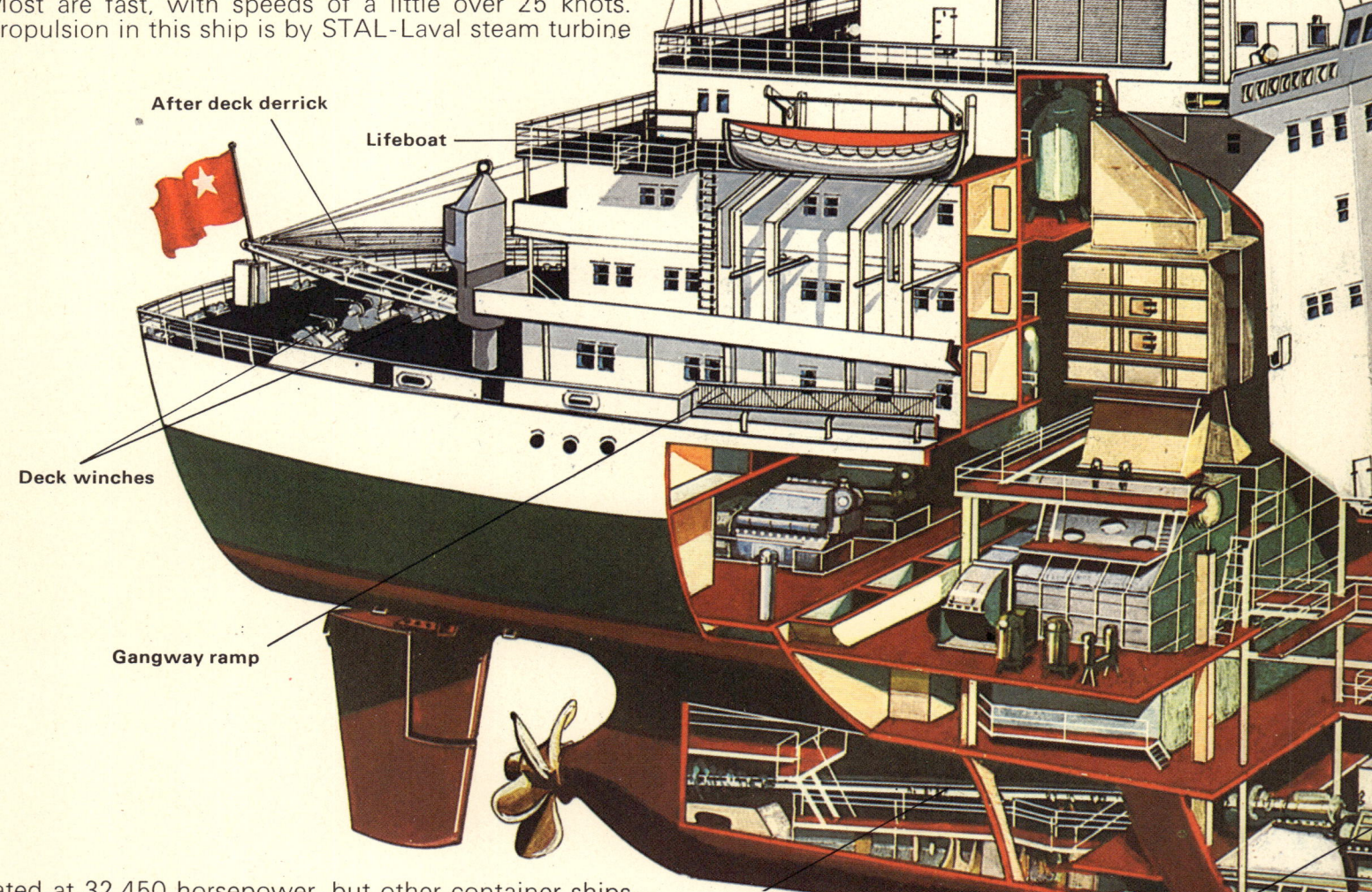

rated at 32,450 horsepower, but other container ships use diesel or gas-turbine engines. Throughout, the object is to increase the machine handling, reduce the manpower on board and on the dockside, and to keep the ship at sea with as little waiting time spent in port as possible.

The loading of a container ship is systematic. The containers are unloaded on the wharf, raised on the gantry and loaded onto the deck of the ship, where they are secured with special clips.

Mobile gantry crane on rails, rapidly unloading cargo

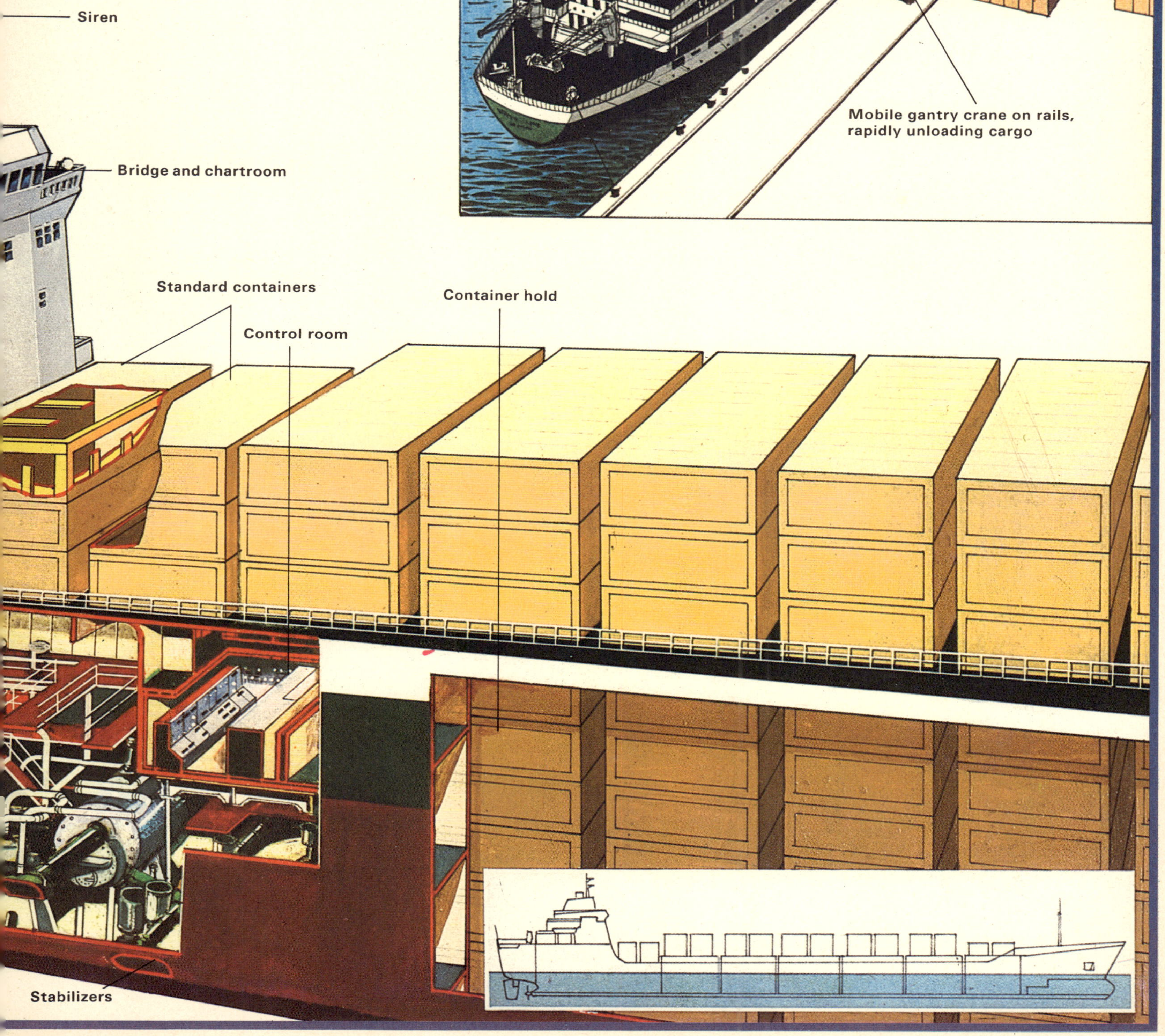

Siren

Bridge and chartroom

Standard containers

Control room

Container hold

Stabilizers

Last of a line: QUEEN ELIZABETH II

The Queens have served the Atlantic run for half a century, and the latest of these – probably the last – is QE 2. She is a dual-purpose ship, designed for cruising and for Atlantic crossings. When she makes the transatlantic run, she carries 2,085 passengers in two classes of accommodation. When cruising she carries only 1,400 passengers. She was built – like her predecessors – at Clydebank, taking two years and two months, and was launched in 1967. Although the smallest of the Queens, she is said to have the highest quality of accommodation.

Her standards of safety are much higher than the law requires. She is divided by watertight doors into 15 compartments controlled from the bridge. She has a safety control room amidships which is constantly manned. A comfortable voyage, even in rough seas, is ensured by her stabilizers, which can reduce a 23 degree roll to a mere 3 degrees. The Decca Navigator is used to collect and compute signals from polar orbiting satellites, providing a 'fix' on the ship's position accurate to within about 200 yards. As a last resort, the ship carries lifeboats and liferafts for 4,000 people although passengers and crew never exceed 3,000.

QE 2 under construction, showing the steel hull and the watertight bulkheads

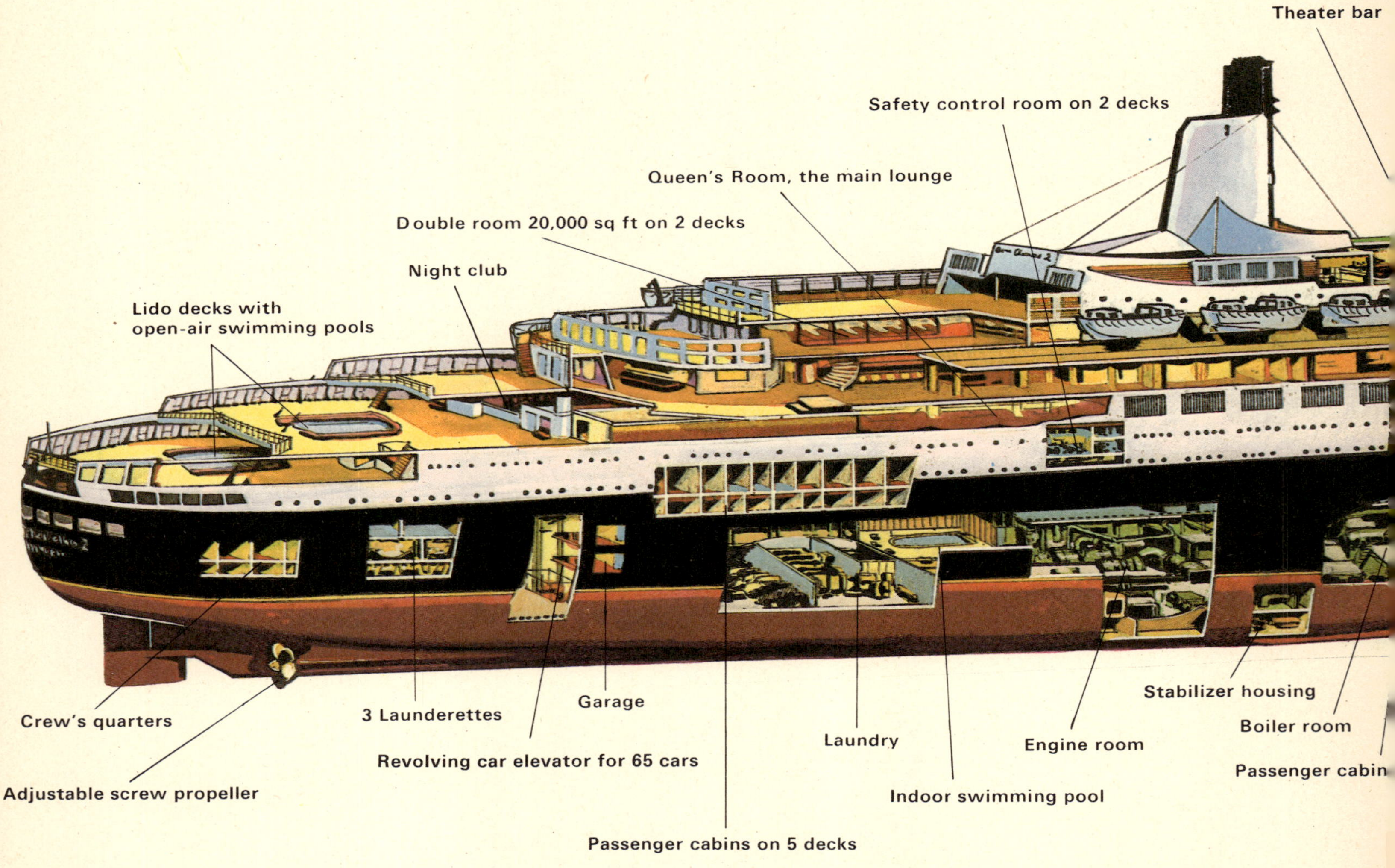

The wheelhouse

Part of the control and communications console in QE 2's wheelhouse

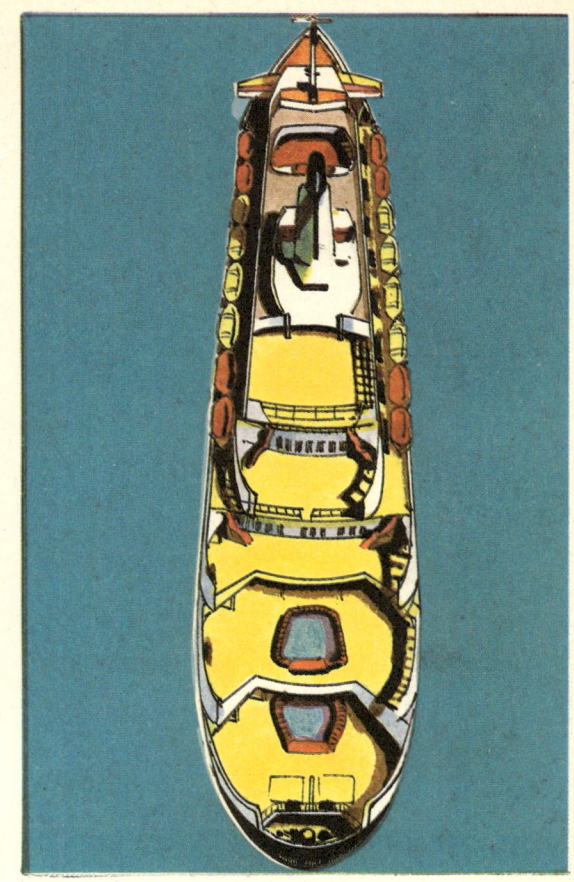
Bird's eye view of the great liner

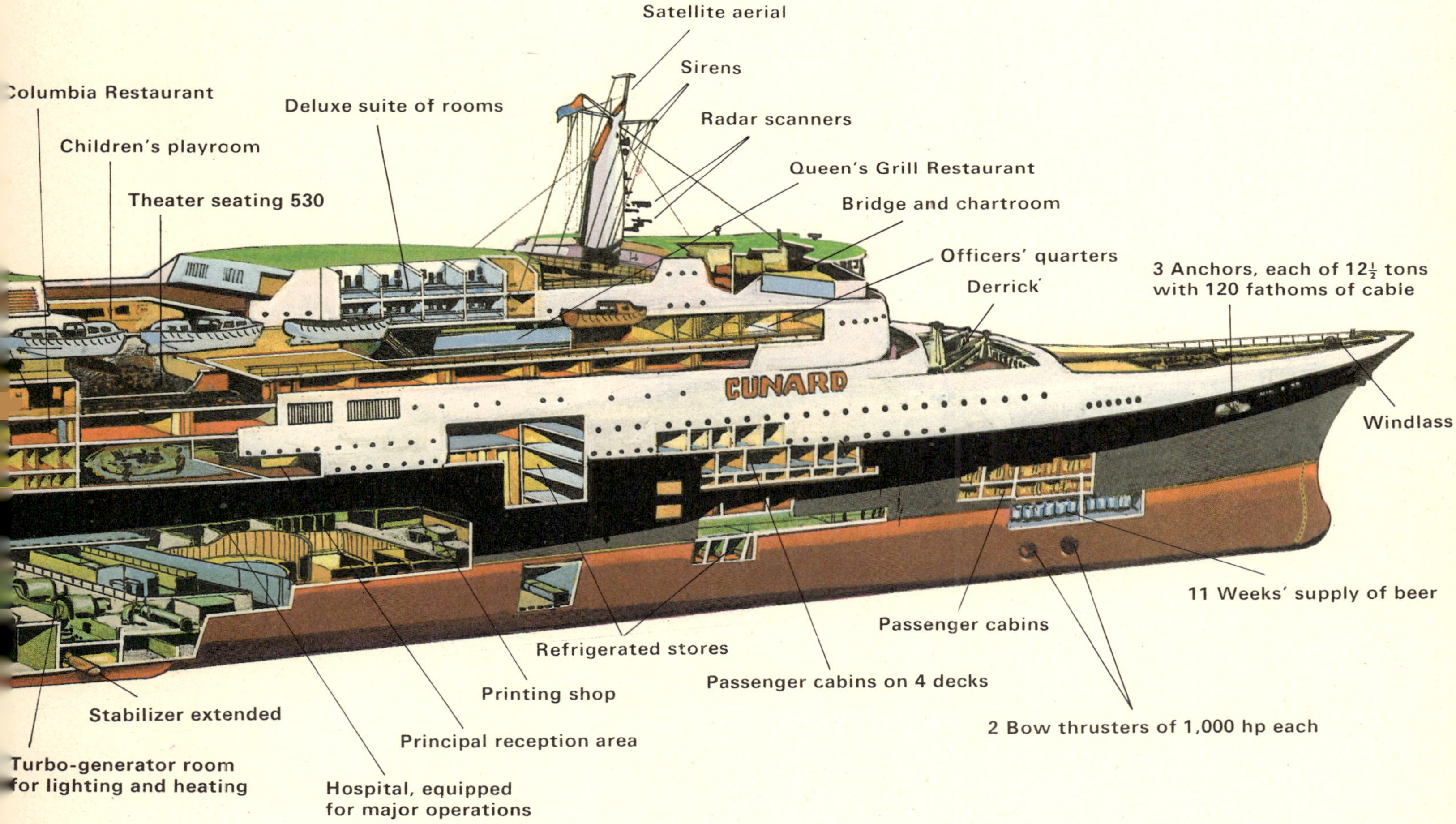

- Columbia Restaurant
- Children's playroom
- Theater seating 530
- Deluxe suite of rooms
- Satellite aerial
- Sirens
- Radar scanners
- Queen's Grill Restaurant
- Bridge and chartroom
- Officers' quarters
- Derrick
- 3 Anchors, each of 12½ tons with 120 fathoms of cable
- Windlass
- 11 Weeks' supply of beer
- Passenger cabins
- Passenger cabins on 4 decks
- 2 Bow thrusters of 1,000 hp each
- Refrigerated stores
- Printing shop
- Principal reception area
- Hospital, equipped for major operations
- Stabilizer extended
- Turbo-generator room for lighting and heating

55

Boat on skis: HYDROFOIL

Most people think of the hydrofoil as a newcomer to the marine world, but in fact commercial services have been in operation for more than 20 years. The hydrofoil is simply a wing that operates in water rather like an aircraft's wing does in the air. It is designed so that water passing over it travels faster than the water that is passing beneath it. The result is a lower pressure above than below. This pressure difference produces an upward thrust that pushes the hydrofoil upwards, lifting the hull out of the water. Since sea water produces about 800 times as much lift as air on a given foil, the hydrofoil does not need to be of such a large area as an aircraft's wing. On the other hand, the hydrofoil must be far more robust to withstand all the buffeting and turbulence of its passage through water. They are usually made of solid metal, often a light strong alloy.

The lift provided by the hydrofoil as it passes through the water lifts the vessel's hull clear of the surface to reduce the drag caused by the friction between the water and the hull's surface. This reduction in drag allows hydrofoil craft to reach higher speeds than would normally be possible with conventional boats.

The hydrofoil is usually propelled by a diesel engine, which is established as a reliable and economical engine for powered craft. The shaft of the diesel engine turns a marine screw propeller, which again is very efficient for speeds up to sixty knots. Other possible means of propulsion include a gas turbine engine driving a water jet, which is the marine equivalent of the aircraft jet engine.

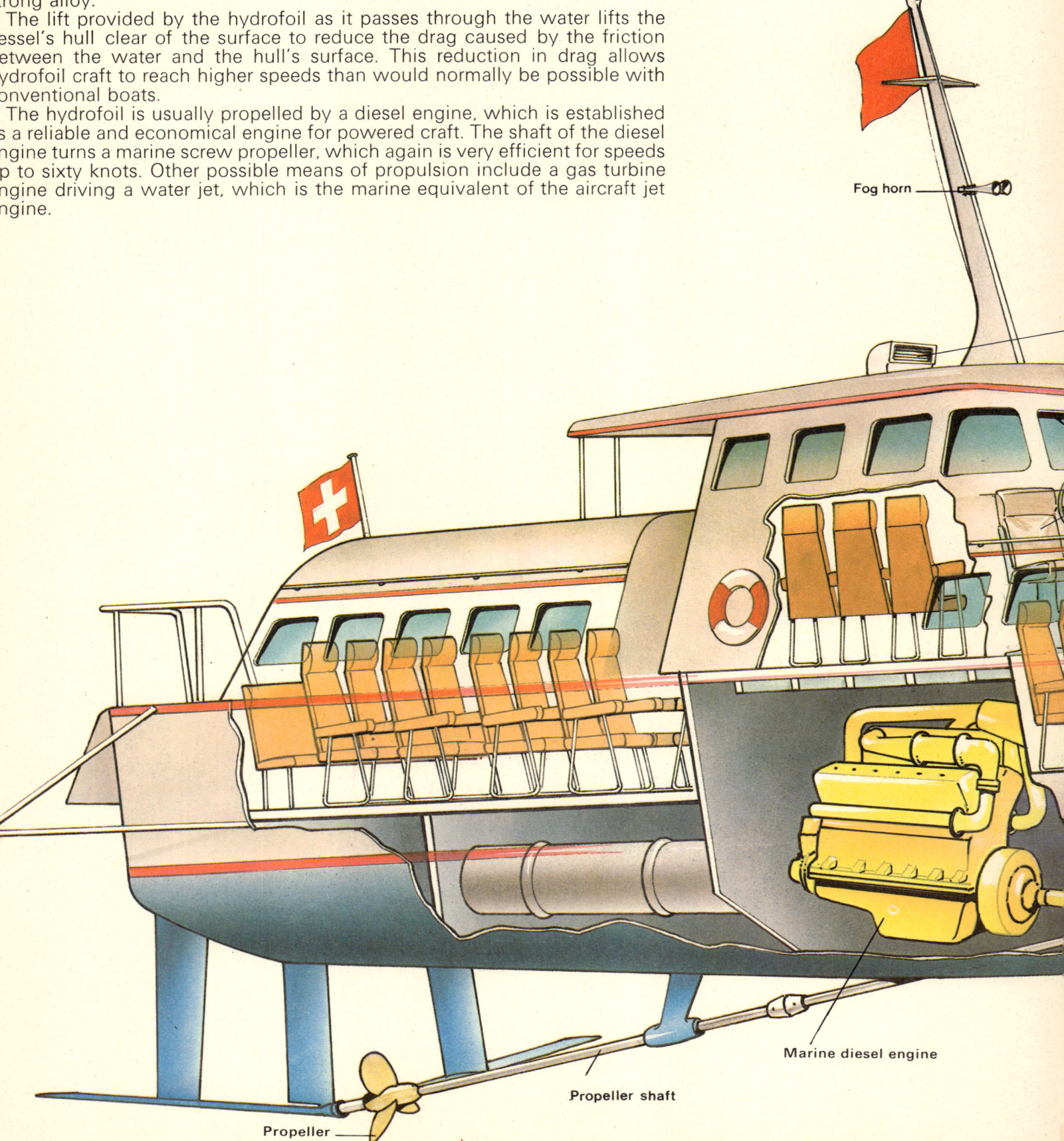

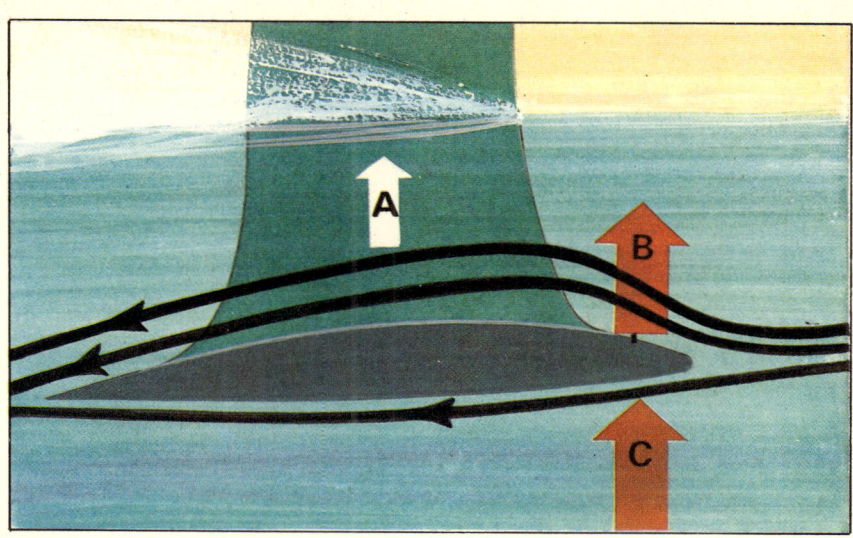

As the foil is forced through the water by the action of the propeller, the water's resistance at C tilts the foil upwards at B, and lifts the hull on its stanchions (A) clear of the surface.

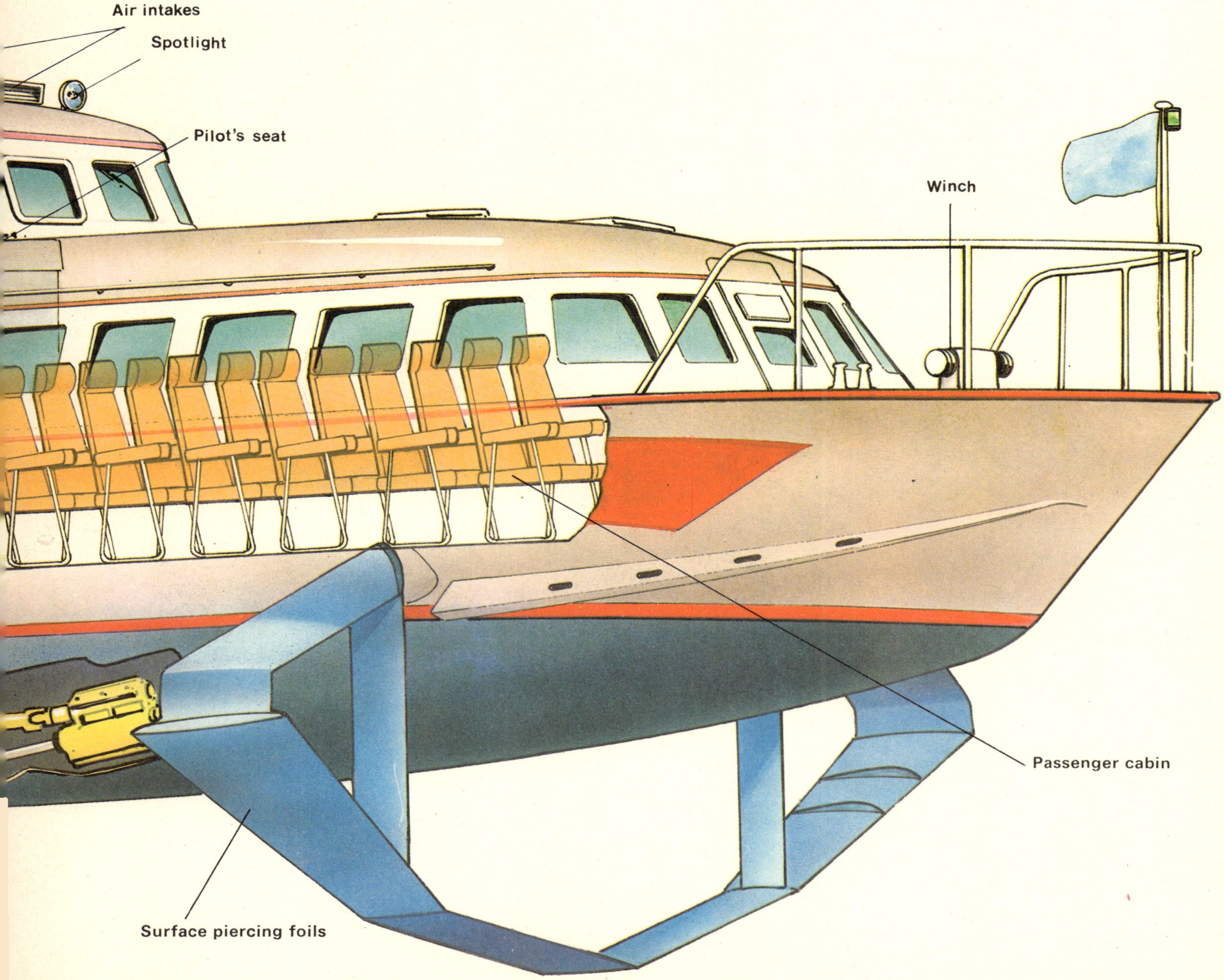

Air intakes
Spotlight
Pilot's seat
Winch
Passenger cabin
Surface piercing foils

Land and sea skimmer: HOVERCRAFT

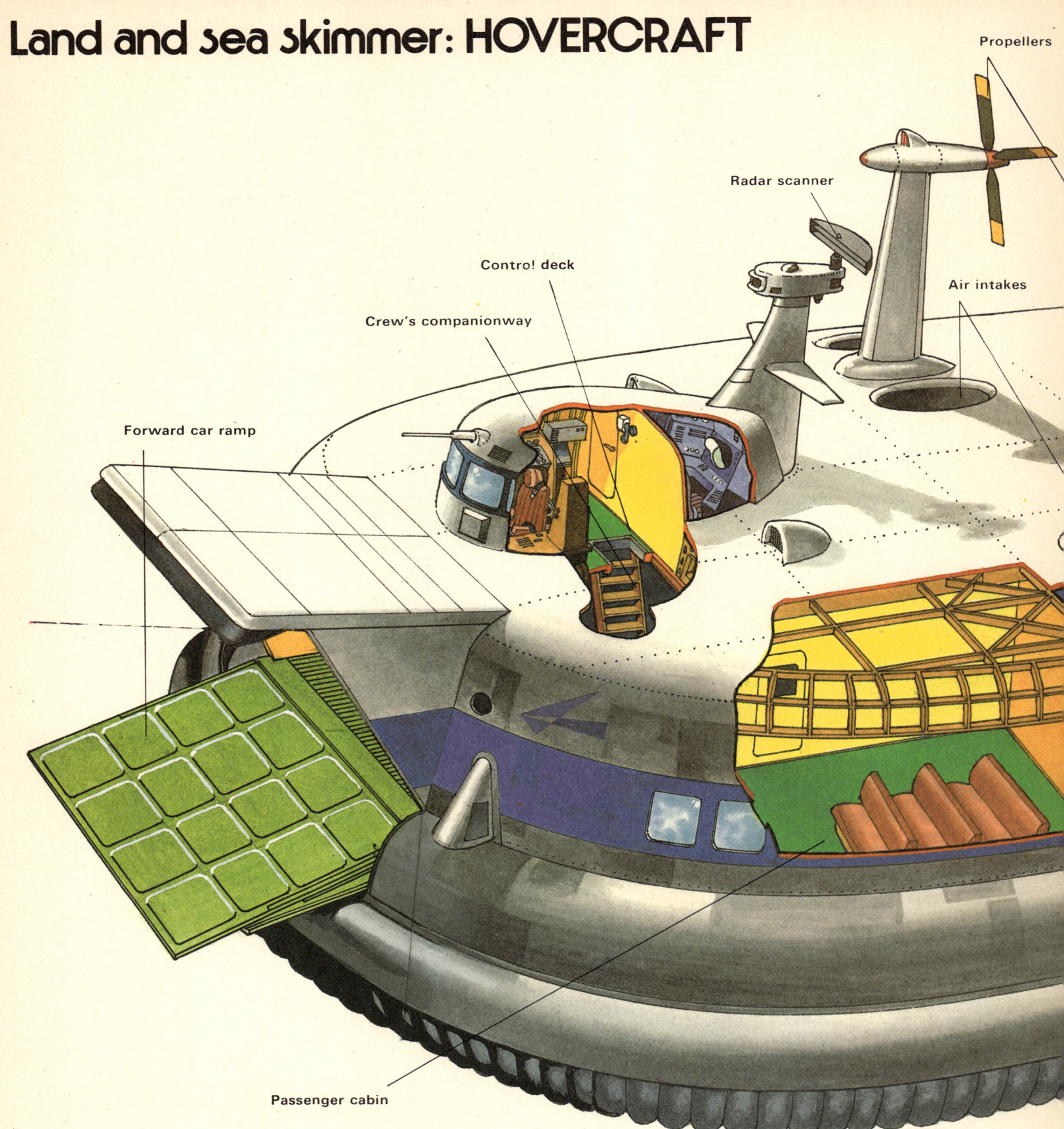

A hovercraft skims over the waves at high speed, throwing up a curtain of spray as it roars along. The flexible skirt that surrounds the craft traps a region of high-pressure air, and the hovercraft is supported on this cushion of air so that it rides just above the surface of the water. For this reason, hovercraft are also known as ACVs – air cushion vehicles – and surface skimmers.

Powerful fans keep the air cushion at high pressure, and propellers usually drive the hovercraft along in the same way as they propel an aircraft. Hovercraft are noisy vehicles, and they pitch about in rough seas as they travel up and down the waves. They are fast compared with boats, though not with hydrofoils – and they do not need to berth at a harbor. A hovercraft can travel as well over flat ground as over water, and so it simply leaves the water and travels up a ramp or over the beach. As the fans stop, the air cushion loses pressure and the craft gently lowers itself to the ground. Some hovercraft do operate from harbors, where they settle on the water and float when at rest. Hovercraft of this type may have a screw projecting beneath the air cushion to provide the power.

The SRN hovercraft is the largest in the world. It weighs 168 tons, and when carrying passengers only will take 600 people. It can also take a mixed load of 34 cars and 174 passengers. It can run at a top speed of 77 knots, but it rarely exceeds its service speed of about 50 knots. The control cabin resembles that of an aircraft, and from it one member of the crew keeps constant watch for boats in the hovercraft's path.

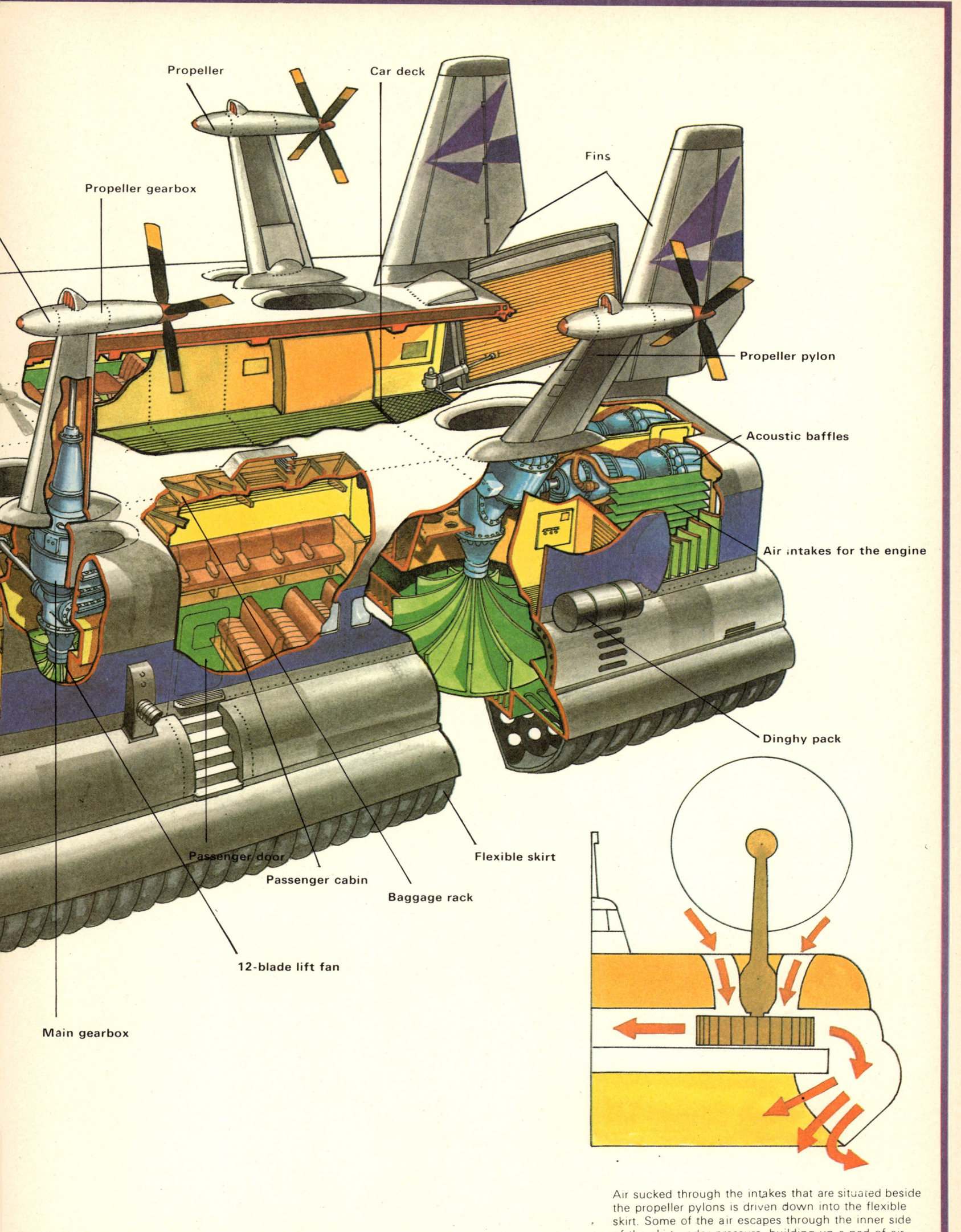

Air sucked through the intakes that are situated beside the propeller pylons is driven down into the flexible skirt. Some of the air escapes through the inner side of the skirt under pressure, building up a pad of air which lifts the craft clear of the water.

Deep sea worker: PISCES

The increased demand for raw materials, and the depletion of many land deposits of minerals, has forced technologists to seek further for their supplies. The ocean bed is rich in many minerals, and has attracted marine geologists and miners who have demanded machines to help them explore and move the deposits. One of these underwater machines is Pisces, which enables a two-man crew to work in a controlled environment much more effectively and safely than divers.

It is necessary in such areas as the North Sea to observe at depths down to 3,500 feet (1,066 meters) and to work down to 600 feet (182 meters). Pisces, with her three viewing ports, makes an excellent observation vehicle, and the crew can lay moorings for drilling rigs, operate valves, bore for rock samples, weld structures and move objects on the sea bed as if they were operating a tractor. The submarine is fitted with two 1,000 watt quartz-iodine lamps which illuminate the depths at which she works. Pisces' telechiric arm acts as a powerful extension of the limbs of its operator working from the safety of the submarine. The crew can stay submerged for up to 60 hours, their air purified by passing it through lithium hydroxide, which absorbs the carbon dioxide from the breathed air. The oxygen is renewed from supplies in oxygen cylinders. The men communicate with their surface support vessel by underwater telephone, closed-circuit TV, and VHF radio; but they are able to make most of their own decisions, using the submarine's instruments, which include compasses, section scanning radar, echo sounder, depth and temperature gauges and ballast indicators.

A larger form of worker submarine becomes operational in 1976. This will be the Laird Sea Bed Vehicle (SBV).

Pisces

Length: 19ft 4in. (6 meters)
Width: 9ft 10in. (3 meters)
Displacement: 24,000lb (10,886kg)
Payload: 2,500lb (1,133kg)
Endurance: 4 hours at 2 knots, 15 hours at ½ knot
Engines: two 3hp motors powered by a 55kwh lead-acid filled battery

Pisces' small size makes it possible to fly her to any place she is needed quickly, but generally she is carried to her site of operations by a support vessel. When she has completed her task, she is winched on board again.

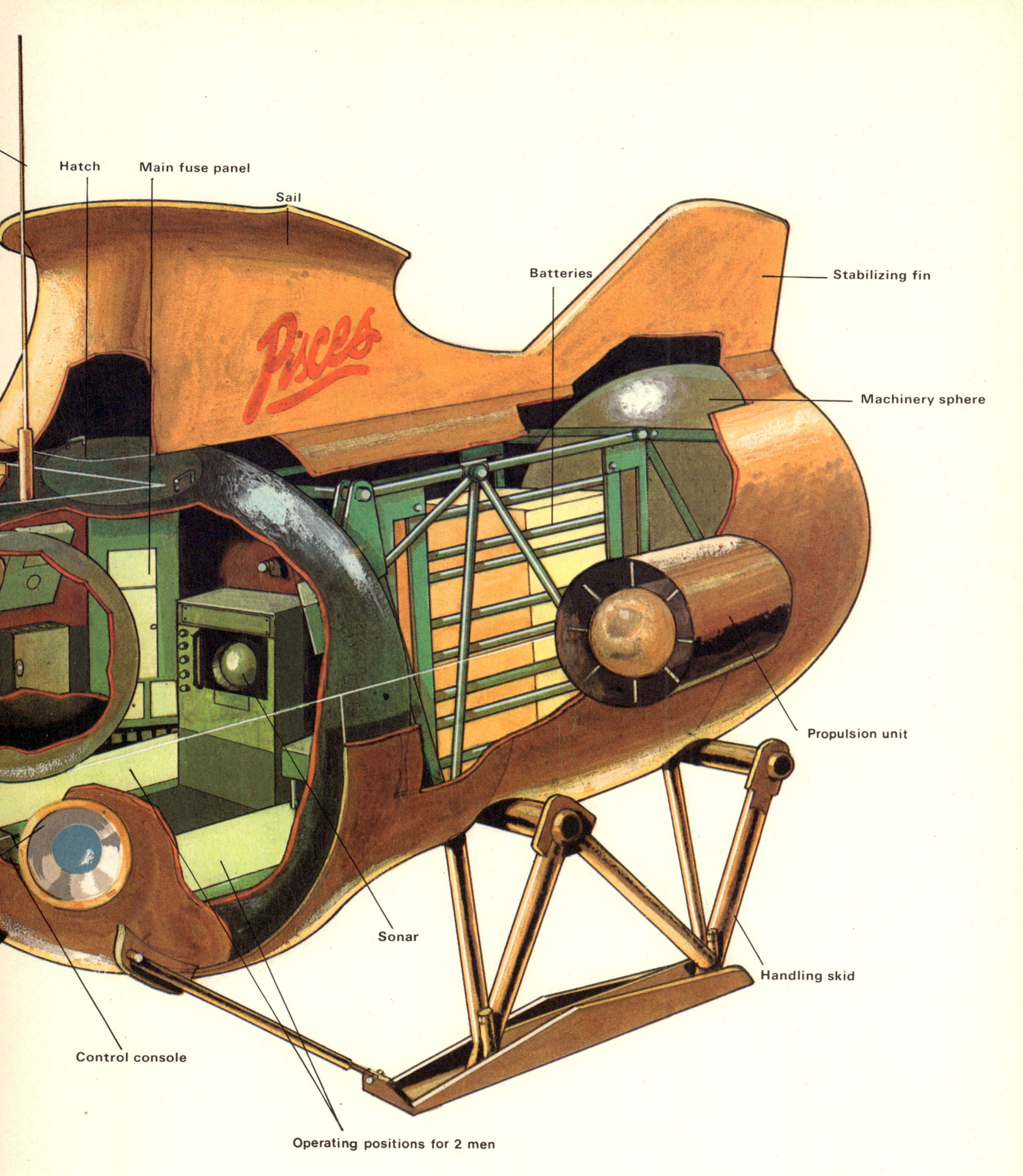

Published in the U.S.A.
by Rand McNally & Company 1976
Library of Congress Catalog Card Number: 76-9940

Acknowledgments
The publisher and author gratefully acknowledge the help received in the form of advice and other generous aid that supported them in the preparation of this book. Of the many who helped, we would especially like to thank: Mr. Neil Ardley; British Rail; Mr. Philip Chapman; Mr. John Clarke; Overseas Containers Ltd; Ruston-Bucyrus Ltd; Mr. Brian M. Service; Townend Thorensen Car Ferries Ltd; Vickers Ltd; and Volvo cars.

Artists
Illustrated by: Norman Barber, Tony Gibbons, Ken Houghton, Brian Lewis, Gary Long, Michael Tregenza, Brian Watson, through Linden Artists Limited.

© **Hampton House Productions Ltd.**
1976
PRINTED IN ITALY